Paul C. Freer

The Elements of Chemistry

Paul C. Freer

The Elements of Chemistry

ISBN/EAN: 9783337275648

Printed in Europe, USA, Canada, Australia, Japan

Cover: Foto ©berggeist007 / pixelio.de

THE

ELEMENTS OF CHEMISTRY

BY

PAUL C. FREER, M.D., PH.D. (MUNICH)
UNIVERSITY OF MICHIGAN

Boston
ALLYN AND BACON
1895

PREFACE.

In undertaking an Elementary Chemistry, I was actuated by a growing conviction that the methods of teaching beginners now very generally in vogue do not make prominent what is *essential* in the science. In the following pages I have endeavored thoroughly to familiarize the pupil with the general aspects of chemical changes, using only a few of the most important elements and compounds for the purposes of illustration; the work is *quantitative*, both in the text and laboratory appendix. Chemistry is growing to be more of an exact science every day, and in it quantitative work can no more be neglected than it can in the study of physics. The atomic theory is not introduced until the pupil has sufficient chemical experience to comprehend its meaning and advantages, and until he thoroughly understands that *theories are based on facts*, not facts on theories. The theory of valence I have only briefly touched upon, as it is not necessary for an understanding of Elementary Chemistry. Its dogmatic application may be productive of more harm than good.

Chemical Equations I have avoided as much as possible, because I wished to give them only the relative importance which belongs to them. The too frequent use of these equations may lead to the view that all reactions which can be so formulated must in reality take place.

The domain of so-called physical chemistry is constantly growing, and can no longer be ignored in works of this kind. For this reason I have introduced some of the simple general facts which have been universally adopted in this branch of the science, notably under the head of electrolysis and double decomposition.

The laboratory experiments are largely quantitative in their nature. Experience has shown me that none of them are too difficult for beginners. It has been too much the custom to seek easy experiments, without considering whether such experiments would teach general principles and, above all, would emphasize quantitative relations. In the study of an exact science, the latter are, however, the most important.

I have endeavored to simplify the apparatus as much as possible, and am sure that the outfit required successfully to follow this book will not be too expensive. Certain kinds of *good* apparatus every chemical laboratory *must* have in order to do successful work; and in all cases where cheap and inefficient things are used, the result is a sacrifice of science in the interests of economy. A good balance, a barometer, two gasometers for the storing of larger quantities of gases, and a combustion furnace are the most expensive pieces required. Whatever is used should be of the best quality, for exact results can only be obtained by exact methods. A chemical experiment, when correctly performed, is as certain to have an unvarying result as is one in physics.

The figures in the Appendix have all been made from

photographs which were taken from apparatus in actual use, so that, if they are used as models, no difficulty will be found in the laboratory work.

I wish to take this opportunity of thanking Mr. Geo. O. Higley for his assistance in compiling the laboratory appendix, and Mr. Geo. A. Bacon for his careful reading of the manuscript copy.

PAUL C. FREER.

ANN ARBOR, *September*, 1895.

TABLE OF CONTENTS.

ELEMENTS OF CHEMISTRY.

CHAPTER I.

CHANGES IN MATTER.

Change in Matter. All matter is subject to change. In some cases, for example when a piece of wood is burning or when water is flowing, the change is perceptible; in other cases it takes place too slowly to be observed. Instances of the latter kind may be found in the wearing away of a rocky mountain along the bed of a stream, or in the gradual decomposition of crystalline rocks into other and different materials through the influence of the weather.

Classification of Changes. It is customary to classify those alterations which are superficial and transitory under the head of physics, and those which involve a change in the essential properties of substance under the head of chemistry. A sharp line, however, cannot be drawn. There is a considerable field lying along the border between the two subjects which is claimed by both. Operations so simple as the dissolving of salt in water, probably involve a chemical change as well as a physical one.

Observation of Changes. From the earliest times, the instability of his surroundings attracted the attention of man, and he speculated as to the causes of the changes going on around him; yet, until the last two hundred years, theories regarding chemical phenomena were not based on scientific observation. The essential difference between the methods of thought preceding the eighteenth century and those of modern times is, that, during the earlier period, theories concerning structural changes in matter were not subjected to proof by experiment. During the later period, theories have been made only to explain facts determined by the most laborious investigations. In consequence of the perverted system of former times, little progress was made toward rationally explaining the composition of even the simplest forms of matter. During the first sixteen centuries of the Christian era, the growth of chemical science was hopelessly retarded, first, by the theory that all substances were composed of four principles, — earth, water, fire, air; and second, by a purely mercenary aim, — the hope of transforming baser metals into gold.

Superficial and Fundamental Differences in Matter. Copper and zinc when melted together form a mass (brass) similar to gold in appearance, and the earlier workers in chemical lines supposed this alloy to be a somewhat modified gold. Copper when heated with arsenic assumes a white color, and the resulting substance was believed to be a metal which differed but little from silver. Thus, merely superficial qualities, such as color, were regarded as essential, while really fundamental differences between various forms of matter were unobserved. We now know that although brass has the color of gold, it nevertheless is entirely different from that metal, for it corrodes when heated, or when brought in contact with acids, while gold does not. Copper and zinc, the constituents of brass, can by the proper means be separated from the alloy; while from gold, no matter what changes the metal may undergo, nothing but gold has ever been obtained. A given volume of gold has an entirely different weight from the same volume of brass. It has a different melting point; in fact, in all the properties which we now recognize as essential, the two substances are distinct.

Aims of Modern Chemistry. It is one of the aims of modern chemistry to distinguish accurately between the various forms of matter, and to discover what changes involve an alteration in the nature of a substance.

The advance from superficial to scientific methods of chemical thought was not brought about abruptly, neither did the idea that experimentation should always be resorted to, as proof of a theory, immediately find adoption.

The Phlogiston Theory. At the beginning of the eighteenth century, a chemist (Stahl) constructed a theory which apparently furnished a uniform and consistent explanation of a common chemical phenomenon, — combustion. According to this theory, any substance capable of burning contained an element, or principle, which was called *phlogiston.** This principle was supposed

* At one time phlogiston was considered as a material substance; at another, as merely a principle devoid of weight. Again, it was identical with the element at present called hydrogen; at another time it was supposed to have negative weight (i.e., to be repelled by the earth). It was also looked upon as the principle of fire. This confusion of ideas is

to pass off while the body was burning, the remainder being "dephlogisticated." Charcoal, for example, burns readily and leaves a small quantity of ash. During the last century this ash was considered a chemical element which, when united with a large amount of phlogiston, formed charcoal. If iron-rust is heated with charcoal, the latter substance disappears, while iron is produced, mixed of course with the ash. The iron-rust was therefore supposed to take up the phlogiston from the charcoal, hence iron was phlogisticated iron-rust. Although this theory was false, nevertheless it led to a great advance in the science of chemistry, because it indicated a direction in which new work could be conducted. During the eighty years following its establishment, greater progress was made than during the preceding fifteen centuries, while at the same time chemistry as an independent science began to be followed for its own sake. Still, investigators of the "phlogiston" period had not entirely outgrown the superficial methods of thought belonging to their predecessors. Although a marked advance had been made, they were still unable, in many cases, to distinguish with sufficient clearness essential from non-essential phenomena. Thus, for example, it was supposed that iron in burning or rusting gave off "phlogiston," and yet the rust weighs more than the iron. True, the ash of the charcoal weighs much less than the charcoal itself; but if a thorough investigation of the combustion of that substance were undertaken, the weight of the ash, plus that of the gaseous products, would be found to exceed that

sure to come where a false theory is called upon to explain a large class of phenomena. Perhaps the nearest approach to "phlogiston" in modern scientific language is to be found in "energy."

of the original charcoal. This essential relation, between the weight of a burning body and that of the bodies produced by burning, was generally ignored during the phlogiston period. It is, however, of such fundamental importance that, when the facts were finally recognized, the theory which overlooked them was of necessity abandoned.* The new one which took its place, being founded solely on experimental evidence, caused a much greater acceleration in the already rapid pace at which chemistry was advancing. In order to understand clearly the reasons for such a radical change of views, it will be necessary to detail one or two important experiments.

Lavoisier's Experiments. A French chemist, Lavoisier, carefully weighed a piece of tin, placed it in a flask, hermetically sealed the latter, and then accurately determined the weight of the whole. He afterward kept the flask heated during a period of some weeks, and observed that the tin had changed into a white powder. At the same time the flask had neither gained nor lost in weight, and consequently no phlogiston could have passed off. On opening the vessel, air rushed in (a partial vacuum having been created), and the total weight was greater than it had been after sealing. Furthermore, the white powder weighed more than the original tin. Lavoisier, therefore, came to the conclusion that the tin, when heated, took up one of the constituents of the enclosed air; that the diminution in volume of the air was connected with the increase in weight of the tin;

* A theory as to the cause of any natural phenomenon must of necessity be abandoned as soon as any fact is discovered which contradicts that theory.

and that the metal did not give off phlogiston when heated. He felt sure that the tin took up a constituent of the atmosphere, but he could not at that time give any absolute proof of his theory, because that portion of the air which is absorbed by burning substances had not as yet been isolated. By a strange coincidence, the chemical theory which was destined to overthrow that of phlogiston was finally established by an ardent adherent of the phlogistic school, Joseph Priestley, who was the first to isolate this unknown constituent of the air. Priestley demonstrated that red precipitate,* when heated, yielded mercury and gave off a new gas, which he called dephlogisticated air,† but which Lavoisier at once recognized as that part of the atmosphere which had acted on the tin. The alteration in the metal was analogous to combustion, and it was Priestley's new gas which united with substances when they burned. Having come to this conclusion, Lavoisier, by repeating Priestley's experiment with red precipitate (taking the precaution to weigh carefully first the substance, and then the resulting gas and mercury produced), further succeeded in discovering a fundamental law of natural science which reads as follows: —

Law of the Conservation of Matter. *During chemical changes the amounts of matter entering into such changes remain constant*, and the total amount of matter contained in the universe never varies. From this it follows that when the oxide of mercury (red precipitate) is decomposed, the sum of the weights of the resulting

* Oxide of mercury.

† Called dephlogisticated air because substances burned in it more readily than in common air. Dephlogisticated air could contain but little phlogiston, since it was capable of taking up phlogiston so readily.

oxygen * and mercury are exactly equal to the weight of the original substance. The tin which Lavoisier heated in a flask gained in weight, but this gain was exactly balanced by the loss in weight of the air enclosed in the vessel. If a candle were burned in a sealed glass globe, the candle would apparently disappear, yet the weight of the globe would remain unaltered. The chemical compounds at first present in the candle, together with a given quantity of oxygen, would have rearranged themselves into other and less complex compounds, differing in character from either the air or the candle, but the total amount of matter would have remained the same.

Importance of Correct Observation. The length of time which it took to come to the conclusion that matter is constant in quantity, and the obstinacy displayed by the adherents of phlogiston, should emphasize for us the importance of careful observation, as well as of a conscientious weighing of all the facts in regard to a given chemical change, before we venture to construct any definite theories regarding it.

Importance of a Knowledge of Facts. In undertaking the study of any science, it is absolutely essential to become acquainted with well-established facts before undertaking to understand any of the theories which have been brought forward to explain the existence of these facts. We must also remember that facts remain as they are, no matter what theories we may advance for their explanation. Almost invariably, however, exactly the reverse of this course has been adopted. As a result,

* The name given to the gas first isolated by Priestley.

students acquire the opinion that chemical science is a mere collection of visionary theories, which experimentation may or may not bear out. This is not the case. When, for example, we speak of atoms, we have at least as exact experimental evidence of their existence as the physicist has of the waves which he discusses in the undulatory theory of light. Because of the danger of confusing really existing things with those which are imaginary, it will be the aim of this book to discuss chiefly facts, leaving the theoretical deductions to be made from them for a larger work.

CHAPTER II.

CHEMICAL NOMENCLATURE.

THE language of every science is peculiar to itself; advances are recorded by alterations, and history is preserved by the use of old expressions. In order to express one's self properly and intelligently, a knowledge of accepted nomenclature is necessary. An extended acquaintance in this line can come only after considerable experience, still it is not out of place to become familiar with a few terms in chemistry even prior to any advance in the study.

Chemical Elements. Toward the end of the last century, the idea became firmly established that those substances which cannot be decomposed into two or more simpler ones, having entirely different properties, are to be considered as chemical elements. This term, however, must not be understood as meaning substances which may not at some future time be broken down. Indeed, past experience indicates that such decompositions will be brought about in substances now classed as elements. An example of such an experience is found in the history of the substance known as caustic potash,* which was formerly supposed to be non-decomposable, and was called an element. Afterwards it was decomposed into a metal (potassium) and two different

* Potassium hydroxide.

gases, oxygen and hydrogen. These three substances are now considered elements; but subsequent experiment may possibly prove these, or any other so-called elements, to be compounds, as was done in the case of caustic potash.

Number of Elements. We are at present acquainted with about seventy different kinds of matter which have not been decomposed into two or more simpler forms.* Five of these are gases at ordinary temperatures and pressures, two are liquids,† and the remainder are solids. These substances are our present chemical elements.

Classification of the Elements. Elements can be roughly divided into two great classes; namely, metals and not-metals. The most pronounced metals are malleable and ductile, are good conductors of electricity, and have a so-called metallic lustre.‡ The most characteristic not-metals, whether gaseous, liquid, or solid, are neither malleable nor ductile, do not conduct electricity (or at least are poorer conductors than the metals), are frequently transparent (oxygen, nitrogen, sulphur crystals) or translucent (bromine, iodine in thin sections), and they have no metallic lustre. It is a matter of experience that the most pronounced metals easily form stable compounds with the not-

* A table of elements will be found at the end of the book.

† A third element, called rubidium, is liquid at about blood heat, and possibly a fourth very rare element, cæsium, may be liquid when it is pure.

‡ An appearance like that of the most familiar polished metals, such as silver, gold, brass, etc., is termed a metallic lustre. Substances with metallic lustre are not transparent or translucent except in the very thinnest plates.

metals, but no absolute prediction as to the behavior of two elements toward each other can be made, even if it is known that one is a metal and the other a not-metal.

Nomenclature of Compounds. Compounds containing but two elements are called binary ones. In the nomenclature of such compounds, the termination of the name of one element is, as a rule, changed to *-ide*, the name of the other remaining unaltered.

Examples: — The compound of iron and oxygen is called iron oxide or oxide of iron.

The compound of hydrogen and sulphur is called hydrogen sulphide or sulphide of hydrogen.

The compound of chlorine and sodium is called sodium chloride or chloride of sodium.

The compound of phosphorus and oxygen is called phosphorus oxide or oxide of phosphorus.

The same element may, however, enter into combination with a series of others. In such an event, it is customary to form for the series a class-name taken from that of the element which all of the members have in common, the termination being altered to *-ide.*

Examples: — Oxides are compounds of oxygen with other elements. Sulphides are compounds of sulphur with other elements. Chlorides are compounds of chlorine with other elements. Phosphides are compounds of phosphorus with other elements.

In naming such a series of compounds, it is invariably the name of the not-metallic element, or of the element acting like a not-metal,* which suffers a change.

* Arsenic, for example, is an element which frequently acts like a not-metal.

EXAMPLES: — Sodium, potassium, and iron are metals; chlorine is a not-metal. The terms sodium chloride, potassium chloride, iron chloride, are, therefore, in use, while the expressions chlorine sodide, potasside, or ironide are inadmissible, although logically just as correct as the others.

Where the binary compound is formed of two not-metals or two metals, then, of course, convention must settle the rules.

EXAMPLES: — We speak of an oxide of sulphur, a sulphide of phosphorus, a chloride of iodine; yet even in such cases it is the name of the most pronounced not-metal which is altered.

Popular Names. It must be remembered that not a few chemical compounds retain the names which were in use before the chemical theories of the present century were adopted, and that some have since that time been given special designations.

EXAMPLES: — Oxide of hydrogen is called water; a nitride of hydrogen is called ammonia; a phosphide of hydrogen is called phosphine.

Chemical Elements and Chemical Compounds. In chemical work we have two conditions in which the elements are encountered. The first is that of the elements in their separate and individual state, the second that of the elements united in compounds. A compound differs radically in behavior and properties from the elements of which it is composed. A chemical compound is not a mechanical mixture (as sand and salt would be); for the mixture has properties which are the mean between those of its constituent parts, while the compound has no resemblance to its components.

CHAPTER III.

WATER.

The Physical Properties of Water. In taking up the subject of chemistry, it is best to study thoroughly a few chemical bodies and the changes which they undergo. We can in this way become familiar with chemical phenomena without introducing theories which are apt to give the impression that chemistry is a visionary and unreal science. Of all known bodies, water is the one which naturally suggests itself for first consideration.

Method of Study. In studying this substance we examine first its physical characteristics, and secondly its chemical characteristics.

Physical Properties of Water. Water is a liquid substance, transparent and nearly colorless when pure, though in very thick masses it apparently has a deep blue color. When heated, it expands; when cooled, it contracts, excepting in the interval of temperatures between 4° and 0°, * [1] where exactly the reverse takes place. At 0° pure water changes to ice, and the ice, following the usual law, continues to contract as the temperature is lowered. At 100° † water boils, and then changes to a vapor, ‡ which expands regularly when further heated.

* Throughout the book temperatures are indicated by the Centigrade thermometer.

[1] The superior numbers in the text refer to the numbers of the laboratory notes in the Appendix.

† At the standard pressure of one atmosphere at the level of the sea. This standard is equal to a pressure of 760 millimetres of the mercury barometer. The pupil can become acquainted with the structure and uses of the barometer and the thermometer, from Note 1 of the Appendix.

‡ A vapor always exists in the presence of the liquid from which it is

Solutions. Water has the power to dissolve a large variety of other substances which are then said to be in solution. Solutions are homogeneous mixtures of two or more substances, at least one of which must be a liquid.*

Solutions of Liquids in Liquids. When all the substances are liquid, then the following special cases must be distinguished: The solution may take place in any proportion if the liquids are perfectly miscible (alcohol and water), or the liquids may only partially mix (ether and water). In either case it is not proper to speak of one liquid as dissolving another, for the alcohol has just as much to do with dissolving the water as the latter has with dissolving the alcohol. Where one liquid only partially mixes with a second, it is also true that the latter will take up an equal proportion of the former. Not infrequently two liquids, when brought in contact, will not mix at all (oil and water).

Solution of Solids in Liquids. When a solid dissolves in a liquid, the latter can take up only a certain quantity called a maximum, which is constant for constant temperatures. In the case of any given solid, this maximum of weight which can be dissolved in a liquid under fixed conditions, is called its solubility,† and the

formed. When the temperature is high enough to vaporize the liquid completely, and to keep it vaporized, we use the term gas instead of vapor.

* Two or more gases can also form homogeneous mixtures; indeed, all gases are perfectly miscible; yet such mixtures, being formed as they are without change in the total volume or in the total heat, can scarcely be compared with ordinary solutions.

† The solubility of a solid is expressed either in percentage by weight of solid which is dissolved in 100 parts of solvent, or in percentage by weight which is dissolved in 100 parts of solution.

solution is said to be "saturated"; i.e., it has taken up of the solid all that it can.

As a rule, the solubility of a solid is increased by increase of temperature. A solution which is saturated at a given temperature will, therefore, be unsaturated at a point above this, and, as a consequence, will take up more of the solid. On the other hand, when a saturated solution cools, a separation must occur, as the solubility of the solid diminishes.*

In an unsaturated solution, when the freezing point is reached, the water † will separate in the form of pure ice until the unfrozen solution becomes saturated; then the solvent and the dissolved solid will separate simultaneously, sometimes in definite relations by weight. The solid mixtures which are formed in this way will have definite compositions, and, for that reason, have sometimes been considered as chemical compounds, and have received the special name of cryohydrates.[2]

Water of Crystallization. Not infrequently substances separate in the form of crystals from solutions in water.‡ These crystals often contain a definite quantity of water for a definite weight of solid. The form of the crystal, as well as the relative amount of water, never varies for the same substance under the same conditions. The water which is so combined to form a solid crystal is called *water of crystallization.* Compounds formed

* Generally the solid separates in a definite geometric form, with plane faces which have a uniform angle with each other. The process is then called crystallization, and the solid so formed is a crystal.

† Provided the solvent is water. The phenomena are the same with any other liquid, provided it be capable of freezing within the limits of the experiment. In the case of a liquid like alcohol, which does not freeze, the cooling of course goes on until the solution is saturated, then separation begins.

‡ The same is true of a number of other solvents.

between a substance and its water of crystallization can be easily broken down. Sometimes the water passes off even at ordinary temperatures, and the substance crumbles to a powder (soda crystals); sometimes the temperature must be raised to the boiling point of water, or even above this, in order to produce the same result (alum crystals). A number of cases are known in which a portion of the water cannot be driven off, even if the temperature is increased far beyond the boiling point; but a careful investigation of these exceptions proves that water is no longer present as such, but has been broken down to enter into an entirely new class of compounds. True water of crystallization is always easily expelled.[3]

Efflorescence and Deliquescence. Substances which lose a part of their water of crystallization at ordinary temperatures are *efflorescent* (soda crystals, copper sulphate crystals). On the other hand, bodies are known which greedily absorb moisture when they are brought in contact with it, sometimes attracting enough liquid to dissolve them. Such bodies are *deliquescent*.[4]

Anhydrous and Hydrated Crystals. It must always be remembered that crystals need not necessarily contain water of crystallization (diamond, chromate of potassium, nitrate of silver, and saltpetre); indeed, by far the greater part do not. Substances which contain no water of crystallization are said to be *anhydrous;* those with it, *hydrated.*

Solutions of Gases in Liquids. There is as great a variation in the solubility of gases as there is in that of solids and liquids, some being very readily dissolved (ammonia, hydrochloric acid); some less readily (chlorine); others with great difficulty (hydrogen). With one or two exceptions, the solubility of gases *diminishes* as the temperature of the solvent *increases*. An increase of the pressure to which a gas is subjected brings with it a corresponding increase in the solubility of the gas. When the pressure is removed, the dissolved gas passes off until the normal solubility at the temperature and pressure of the atmosphere is reached.* The solubility of a gas varies directly as the pressure.[5]

Impurities in Water. Distillation. The impurities found in water are either mechanically suspended or dissolved. The suspended impurities can be removed by filtration,† the dissolved ones, in certain cases, by distillation. Distillation consists in heating the solvent to its boiling point in an apparatus suitably arranged for collecting and condensing the vapors in a separate vessel. Where the substances dissolved are solid, they are in most cases ‡ removed by this process. Liquid substances dissolved in water can be removed by distillation only when the boiling point of the liquid is much above or below that of water; for, unless this be the case, a portion of one liquid will always vaporize with the other. Even where there is a considerable interval between the boiling points, the operation of distilling must be re-

* The effervescing of soda-water is an illustration of this phenomenon. The gas, carbon dioxide, is forced to dissolve in the water by great pressure. When the pressure is removed, the carbon dioxide must escape.

† By passing the water through a porous substance; for instance, especially prepared blotting-paper or unglazed porcelain.

‡ Certain solid substances vaporize with the water. Such substances cannot, of course, be removed by ordinary distillation.

peated several times, the lower and higher boiling portions being collected, separated, and then again distilled from separate vessels. This operation is called *fractional distillation*.[6]

Natural Waters. Natural waters are all more or less impure. Rain-water or melted snow is nearest to absolute purity, but even this contains solid and gaseous substances collected in passing through the atmosphere. When rain-water soaks through the soil, it takes up a certain portion of the solid substances with which it comes in contact. If poisonous substances in the soil or organisms deleterious to life are taken up, the water will be unfit for drinking purposes; and this is sometimes the case in well or spring water. Where certain constituents of the soil are of an extremely soluble nature, like common salt, Epsom salt, etc., they are of course taken up by the water which filters through. Where such a water escapes from the soil, it produces mineral springs. Naturally the impurities dissolved in well or spring water vary with the nature of the soil through which the rain-water forming the well or spring has trickled.

CHAPTER IV.

WATER (*Continued*).

THE first inquiry into the chemical nature of water should have for its object the definite solution of the question whether water is a chemical element or a compound,* and in order to find an answer we must resort to experiment.† At the same time, it will be possible, by the study of a few facts, to get a clearer conception of what is actually meant by a chemical element.

Action of Sodium and Potassium in Water. If a piece of the metal sodium is placed in contact with water, an instantaneous change takes place; the sodium becomes hot, it melts, and the globule of metal will move rapidly round on the surface of the water.‡ If the water is thickened with starch paste so that this movement cannot take place, the heat developed will finally pro-

* See page 8.

† In former times water was regarded as an element. One reason for this was that water is so common, and it appears to be generated during so many chemical processes (for instance, the combustion or distillation of wood). The chief means for decomposing substances which the older chemists had at their disposal was the action of fire. Certainly no temperature which they could produce would destroy water, and this was another reason for supposing it to be an element. We should ourselves be compelled to regard water as an element *were we not able to decompose it into two different substances.*

‡ The specific gravity of sodium is less than that of water; that is, a given bulk of water would weigh more than the same bulk of sodium. The sodium, therefore, floats on water.

duce a flame which is caused by the burning of a gas which is passing off. That such a gas is really liberated is easily proved by placing the piece of sodium in a wire net, and inverting over this a tube closed at one end, and filled completely with water. The gas will rise in bubbles, and, if the original quantity of the sodium was sufficient, will entirely fill the tube. If the latter is now removed, mouth downward, and held over a flame, the gas will take fire, and burn up completely when the tube is inverted. Essentially similar phenomena will be observed if the metal potassium * is substituted for sodium, except that the change is much more violent. The gas bursts into flame even when the water has not been previously thickened with starch paste. The color of the flame with sodium is yellow, with potassium bluish-violet; and to a superficial observer it might appear that the gas given off by potassium and water is different from that produced by sodium and water. That this is not the case, however, can be shown by comparing the flame of the gas collected in a test-tube from sodium with that collected from potassium. In both cases the flame will be nearly colorless.[7]

The yellow and the violet colors are due to something derived from the potassium and the sodium. The explanation is that small particles of the metals are carried up into the flame, and, being heated to incandescence, give off the colored light peculiar to each.

* Potassium is a metal very much like sodium in appearance and character. Its specific gravity is less than that of water, therefore it floats. Both sodium and potassium are soft, can be easily cut with a knife, and the freshly cut surfaces possess a brilliant metallic lustre, which almost instantly disappears on exposure to the air. Because they are so easily altered by exposure to the air, they must be preserved under petroleum.

Nature of the Action of Sodium and Potassium on Water. Is the gas which is developed in this process due to a breaking down of the water, or does it come from the potassium and the sodium? Either view seems plausible; and, indeed, had the two metals in question been known in the last century,* the latter interpretation of the change would undoubtedly have been the one adopted. Sodium and potassium would have been regarded as *compound substances*, and it would have been supposed that they gave up a gas (phlogiston) when brought into contact with water. The substances, therefore, which remained behind, dissolved in the water (caustic soda and caustic potash), would have been classed as *elements*. Such an explanation is entirely incorrect, however, for the gas does not come from the potassium or the sodium. It results from the breaking down of water into two parts, one of which passes off as a gas, while the other unites with the potassium or sodium to form a *new compound* body, which dissolves in the water. That this is really the case is shown by the following experiments: —

Decomposition of Water by the Electric Current. If water † is placed in a cup, into the bottom of which two pieces of platinum foil are placed in such a manner that they can be connected with the two poles of an electric battery, and an electric current is passed through the water, bubbles of gas will appear at both pieces of foil. If a tube, closed at one end and filled

* Sodium was discovered in 1807 by Davy; potassium, in 1808, by the same investigator.

† Rendered slightly acid by means of sulphuric acid; pure water does not conduct electricity.

with water, is inverted over each piece of foil, the gases can easily be collected. It will then be found that the water in one tube is expelled more rapidly than in the other. If the tubes selected are of the same size, one will be exactly filled with gas at the moment when the other is only one-half full.[8] It is evident, then, that water decomposes into two gases, one of which separates at the positive, the other at the negative pole; and it is also true that there is about twice as much of one gas as of the other. If the tube containing the greater quantity * is removed, and brought in contact with a flame, the gas will be found to burn with a flame identical in appearance with the one observed during the combustion of the gas produced from sodium and potassium. It is probable, therefore, that the gas liberated when experimenting with those metals had its origin in the *water* with which they were brought in contact, and that it did not come from the metals themselves.†

The proof that the gas which is formed by the action of sodium or potassium on water does not come from the metal, but that it has its origin in the water which is decomposed, has also been obtained by other methods. These, however, would involve more difficult experimentation than is possible for beginners. It has been

* Experimentation will show that this has been collected at the *negative* pole of the battery. It is, therefore, the electro-positive constituent.

† The proof that the gases obtained by the action of sodium or potassium on water, and the gas collecting at the negative pole when water is decomposed by the electric current, are *really* identical has not been shown by the experiments which have been given. The only point made use of is that they both burn with a similar flame. Something more than this is necessary before we can really be satisfied as to their identity. This proof is found in the fact that the gas from both sources, when burned, forms the same substance; namely, water (see page 27).

shown that the sum of the weights of the sodium used and of the water decomposed is equal to the sum of the weights of the gas produced and of the compound of sodium which remains dissolved. If all excess of water is removed by heat, then the sodium compound remaining weighs *more* than did the sodium originally used. The sodium, therefore, could not have *given off the gas*, for in that event it would have *diminished* in weight.

That the volume of one of the gases into which water is broken down is *exactly* twice as great as that of the other gas cannot be proved from the decomposition of water by the electric current, since the method employed is not sufficiently accurate. A more careful and extended study of the composition of water by other means is necessary to show that the ratio of the two gases by volume is exactly 2 : 1.

Sodium and Potassium are Elements; Water is a Compound. A number of observations similar to the ones just given, and involving the action of sodium and potassium on other substances which are chemically like water, have brought us to the conclusion that sodium and potassium are *not decomposable.* In all cases of experiment, when a theory that these metals are decomposed into simpler substances might seem plausible, closer analysis will show that it is not sodium or potassium that is broken down, but *the other substances* with which these metals are brought in contact. Sodium and potassium are, therefore, *elements*, and water is a *compound.* Having settled this, the next step is to discover what and how many dissimilar substances can be produced by the separation of the water into its constituent parts.

Hydrogen and Oxygen. We have seen that when an electric current is passed through water, a gas separates both at the positive and at the negative pole (see page 22). The accumulation at the latter is twice as great as that at the former, and the question at once arises: are the gases identical, or do they differ from each other? To all appearances they are alike, both are colorless, neither possesses an odor; yet a distinction will at once become apparent if a lighted taper is applied to each in turn. The one which has been collected at the *negative* pole burns with a flame identical in appearance with that shown by the gas obtained from the action of sodium or potassium on water. Indeed, as we have seen, it is in reality the same substance. This gas is termed hydrogen.* The gas collected at the *positive* pole does not burn, but it causes a taper placed in it to burn much more energetically than it would in the air. Even if the taper has but a feeble spark, it will burst into flame. This gas is called oxygen.[9]

Formation of Water from Hydrogen and Oxygen. It has been shown that hydrogen and oxygen can be obtained from water; can water also be produced from these substances? Because if it can, then these two elements alone make up the compound, water. The following experiment will demonstrate this fact: —

A tube, sealed at one end, is filled with mercury, then

* Neither of the methods which have been discussed are expedient for the preparation of the considerable quantities of hydrogen necessary for ordinary laboratory purposes. To accomplish this end another way is necessary, which is described in Experiment 9, Appendix. Both oxygen and hydrogen are called *elements;* for no experiments have been discovered by means of which these forms of matter can be separated into two or more different substances as, for example, has been done in the case of water.

closed at the other end by the thumb, and inverted over a trough filled with the same liquid. Perfectly pure hydrogen is then run into the tube until the latter is filled for about five centimetres.* Oxygen is added so that its bulk is approximately two centimetres, and the tube is firmly closed. The tube used in this experiment has two short platinum wires fused through its sides near the closed tip. These are connected with the two short poles of an induction coil, and a spark is allowed to pass through the mixture of gases. A bright flash is seen, a slight explosion is heard, and the mercury rises in the tube. A portion of the gas will remain, and this will be found to have the properties of hydrogen. All the oxygen with a part of the hydrogen has been used to form water. If the tube is carefully marked in cubic centimetres, so that the volumes of the gases which were introduced can be accurately measured, it can easily be proved that the contraction in total volume is such that exactly *two volumes of hydrogen* have united with *one of oxygen* to form water.[10] †

A mixture of hydrogen and oxygen remains without change indefinitely under ordinary conditions, union only taking place if a lighted taper is brought in contact with the gases, or if an electric spark is passed through the mixture. An explosion which may be violent will then result, provided the proportions in which the gases taken are nearly two parts by volume of hydrogen to one of oxygen.

* For the preparation of pure hydrogen, see Experiment 9, Appendix.

† Of course the amount of water which can be formed in a small glass tube is too little for identification as such. In order to prove that water is really the product of the action of hydrogen on oxygen, a large quantity of hydrogen must be burned in oxygen. In order to be able to furnish this proof, the pupils should perform the experiment given in Note 12 of the Laboratory Appendix.

Relation between the Volumes of Hydrogen and Oxygen, and of the Water Produced. If exactly two volumes of hydrogen and one of oxygen are exploded in the apparatus described in the preceding paragraph,* the mercury will completely fill the tube, because the volume of liquid water which is formed is inconsiderable, compared with the volume of the gases from which it has been produced. But if the tube which contained the hydrogen and oxygen is kept heated by a steam-jacket, so that its temperature is that of the boiling point of water, then the water vapor produced will not only occupy a visible portion, but this volume will be exactly two-thirds of that taken up by the gases before explosion.[11] The facts which have been discovered are, therefore, as follows: Hydrogen and oxygen unite chemically to form a new substance, which is water. This substance is produced by the interaction of exactly two volumes of hydrogen and one of oxygen. The volume of the vapor of this compound is exactly two-thirds of the sum of the volumes of the original gases from which it is formed. If more than two volumes of hydrogen to one of oxygen have been mixed before the explosion, then some hydrogen will remain unaltered; if more than one-half as much oxygen as hydrogen by volume was present, then some oxygen will be found in the tube after the explosion. No matter in what proportions hydrogen and oxygen are mixed, they will unite to produce water only in the proportion, two of hydrogen to one of oxygen.

Relation of Hydrogen to Oxygen by Weight. Obviously a given volume of hydrogen must always have the same

* For an exact description of this apparatus, see Note 10, Laboratory Appendix.

weight no matter where it is found, provided the conditions of temperature and barometric pressure remain identical,* and the same must also be true of a given volume of oxygen. Hence, it follows from the above experiments that *water is always produced by the interaction of hydrogen and oxygen in a certain definite ratio by weight.* This ratio can be discovered by weighing a glass globe which has been emptied by means of an air-pump, weighing the same after filling it with pure hydrogen, then removing the hydrogen, filling the globe with oxygen, and again weighing. Comparing the results, we shall have the relation existing between the weight of a given volume of hydrogen, and that of the same volume of oxygen. This relation has been found to be as *one* to *sixteen;* or,† in other words, if the specific gravity of hydrogen is placed at unity, the specific gravity of oxygen is sixteen. But we have learned that two volumes of hydrogen unite with one of oxygen to form water; hence, the relationship by weight between the hydrogen and oxygen which form water is as *two* to *sixteen*, or *one* to *eight.*‡

One fact still remains to be shown, for we have not yet definitely proved that the product of the explosion between hydrogen and oxygen really is water. The quantity of the latter substance obtainable by the method we have employed is so small that it could not practically be identified. However, all doubts on this subject can be removed by arranging a hydrogen

* See pages 67, 68, 69.

† I.e., if the volume of the globe was such as to hold exactly one gram of hydrogen under the conditions of the experiment, then the oxygen would weigh sixteen grams. The same ratio would be preserved, no matter what was the weight of the original volume of hydrogen.

‡ Throughout this book proportions are given by weight unless volume is expressly stated.

apparatus which will deliver pure, dry hydrogen through a jet,* lighting the gas, and then plunging the jet into a jar filled with oxygen. It will continue to burn more energetically than in the air, and drops of water produced by the combustion will soon collect on the sides of the vessel.[12] Experiment has shown that, no matter what the origin of the hydrogen may be, it always produces water whenever it is burned in oxygen.

Summary. The facts which we have now learned in regard to the chemical nature of water can be summed up as follows: —

1. When sodium or potassium are brought in contact with water, a gas, hydrogen, is given off from the water.

2. Considerable heat is developed during the process, the sodium melting, and, if kept stationary, even setting fire to the hydrogen, while the gas passing from the potassium takes fire even if the metal is moving.

3. Water can be decomposed by the electric current, hydrogen and a second gas, oxygen, being produced by the process.

4. The ratio by volume between these gases when they are formed from water is as two volumes of hydrogen to one of oxygen.

5. Water can be produced from hydrogen and oxygen by exploding a mixture of the two gases.

6. Water is the result of the union of exactly two volumes of hydrogen and one volume of oxygen.

7. Water is formed by the union of *two* parts by weight of hydrogen with every *sixteen* of oxygen.†

8. *Hydrogen being selected as the standard*, the specific gravity of oxygen is *sixteen*.

* See Experiment 9, Appendix. † In exact numbers, 2 : 15.88.

CHAPTER V.

THE PRODUCTS OF THE ACTION OF SODIUM AND POTASSIUM ON WATER.

Caustic Soda and Caustic Potash. When sodium or potassium is brought in contact with water, we have seen that both metals dissolve, while, at the same time, heat is given off. We have further learned that the hydrogen which is produced is a *constituent of the water.* What becomes of the sodium or potassium? Apparently they have disappeared, and the water feels soapy. If the liquid in which the metals were dissolved is evaporated, there will remain, in each case, a white, semi-solid deliquescent * mass, which will readily harden when heated to about 150° centigrade. Owing to the fact that these substances will burn the skin and mucous membrane, and that they contain either sodium or potassium, they are called caustic soda and potash, respectively.[13]

Electrolysis of Caustic Soda or Caustic Potash. Put one of these solids into a small platinum dish, mix with it a little water,† and connect the mass with the negative pole of a powerful battery, while the dish is joined to the positive. The liquid will become warm, bubbles of gas will escape, and globules of a metallic substance will appear at the top of the wire. These can be scraped

* See page 16.

† Not enough to dissolve.

off. They corrode, and even may take fire in moist air. They decompose water, yielding hydrogen. In short, they possess all of the properties belonging to sodium or potassium. We have thus proved that when sodium or potassium dissolves in water, there remains in solution a compound which, on electrolysis, yields sodium or potassium in a manner exactly parallel to the formation of hydrogen from water.

(In the following paragraphs, the nature of caustic potash alone will be considered, the pupil bearing in mind that what applies to that substance is equally true of caustic soda.)

Chemical Structure of Caustic Potash. If caustic potash is heated with potassium,* *hydrogen* will soon pass off. If the hydrogen developed by dissolving a weighed quantity of potassium in water were measured, and the resulting volume were compared with that obtained from heating the same weight of *potassium* with caustic potash, the two quantities would be found to be identical.[14] The action of potassium on water can, therefore, be divided into two stages, — first, one-half of the hydrogen contained in the water is expelled,† leaving caustic potash (*potassium hydroxide*); and then the second half can be removed in the manner mentioned above, leaving *oxide of potassium.*‡

* In a silver dish or glass tube; hot caustic potash attacks a platinum vessel, while the clumsiness and weight of an iron one are objectionable.

† Provided enough potassium is used to change the entire amount of any given quantity of water into potassium hydroxide.

‡ That the oxide of potassium must remain will be seen by a little reflection; for water, as has been proven, is composed of hydrogen and oxygen. The hydrogen in the above experiment has been expelled by potassium; but as no oxygen has passed off, that element must have remained behind in combination with the potassium. That the oxide of potassium is really formed in this way can be proved experimentally by burning a little piece of potassium in oxygen. The oxide so produced is identical with that formed by the complete action of potassium on water.

Formulation of the Action of Potassium on Water. In forming water, two volumes of hydrogen unite with one volume of oxygen. The preceding experiments also show that the hydrogen in water can be divided into two parts. These two facts can be made clearer if we represent the two halves of hydrogen obtained from water each by H, and the oxygen formed by the same means by O. The structure of water will then be represented by the combination HOH; and designating the potassium which acts upon water by a letter K, the two stages of the reaction could be represented as follows: —

$$HOH + K = H + KOH.$$
$$KOH + K = H + KOK.*$$

The potassium in potassium oxide is therefore composed of *two halves*, just as much as is the hydrogen in water, so that caustic potash (potassium hydroxide) is simply water in which potassium has taken the place of *one-half the hydrogen*, while potassium oxide is water in which potassium has replaced *all of the hydrogen.* Such substitutions are by no means uncommon in chemical reactions, and the changes studied above will serve as an example for all others.

The conclusion at which we have arrived, using potassium, would have been equally apparent had we employed sodium.

* In these equations the quantity of water represented by HOH is such that it is *entirely* decomposed by the amount of potassium represented by K. The letter K is used to designate potassium, because it is the first letter of the Latin name (*Kalium*) of that element.

CHAPTER VI.

CHANGES OF ENERGY

Which take place in the Formation of Water, Potassium Hydroxide, and Potassium Oxide.

Work and Energy. Potential Energy. When a substance, either by reason of its position or of its motion, is capable of performing work, it is said to possess energy. When energy is applied to overcoming resistance, we say work is done. The energy possessed by a body may be divided into two classes,—energy of position (potential energy), and energy of motion (kinetic energy). A stone at the top of a hill possesses *potential energy;* for, by reason of its position, it is capable of performing work when the force holding it in place is removed. The work which it can do is measured by the mass of the stone multiplied by the distance through which it acts ($L = MS$).

Kinetic Energy. The capacity for work which a body in motion possesses by reason of that motion is called its kinetic energy; and the measure of this is one-half the mass of the body multiplied by the square of its velocity $\left(\frac{Mv^2}{2}\right)$.*

Conversion of Potential into Kinetic Energy. Conservation of Energy. If a portion of the potential energy pos-

* For the derivation of this formula, consult any elementary text-book on physics.

sessed by a body is converted into kinetic energy by its assuming motion, then the sum of the capacity for work still remaining, and of the kinetic energy already produced, is equal to the potential energy originally contained in the body. This sum is also equal to the kinetic energy which would be produced if its entire potential energy had been used for the performance of work. The sum of the potential and of the kinetic energy is, therefore, constant. (Principle of the conservation of energy.)

If the phenomena attendant upon the burning of hydrogen in oxygen are recalled, it will be remembered that heat is evolved during the process. If this heat were properly applied (as, for instance, in changing water into steam with which to move an engine), it would be capable of performing work.* Hydrogen and oxygen, when in contact, possess a form of energy (chemical energy) which is akin to potential energy. The only distinction between these is that in the latter the capacity for work can be resolved into two factors, the weight of the acting body and the distance through which it acts, while in the former the distances through which the action takes place, and the weights of the particles into which we suppose all bodies divided, are so small that chemical energy cannot be resolved into two factors. The amount of chemical energy can, however, be measured by *the amount of work which given weights* of the interacting bodies are

* This tendency which certain elements have to unite with each other to form compounds is termed "chemical affinity," or "chemism." The nature of the force acting between the elements we cannot explain; but it is by reason of this mutual attraction that they can perform work, just as the stone which possesses potential energy can perform work owing to the attraction of gravitation.

capable of performing. When hydrogen and oxygen unite to form water, their chemical energy is converted into kinetic energy, and is manifested in the form of heat which is susceptible of measurement.

The Energy required to Decompose a Body is equal to that given off in its Formation. A stone when elevated above the ground possesses potential energy. When the force sustaining this stone is removed, it falls, and in so doing changes its potential energy into kinetic energy. To raise it to its original position after it has fallen, we must add exactly as much energy as was given off during the fall. In the formation and decomposition of water we are confronted by a parallel case; for *exactly as much kinetic energy must be used in decomposing a given quantity of water into its constituent elements, oxygen and hydrogen, as was given off in the formation of the same weight of liquid.* We accomlished the decomposition by the electric current, and the quantity of electricity used could perform exactly as much work as could the heat given off in forming the water which was decomposed. This conclusion can be summed up as follows: —

The kinetic energy given off in the formation of a given weight of water is equal to the kinetic energy necessary for its decomposition; or, as the electricity used in the breaking down can be converted into heat,*

* In the case of water it is not practically possible to decompose the compound by heat alone, the temperature at which the decomposition takes place being too high. In order to effect the decomposition we must resort to the electric current. Heat can, however, be used to break down many other compounds; and in such cases it is, of course, possible to measure the heat of decomposition.

we say the heat of formation of a given weight of water is equal to the heat of decomposition.

Energy Changes in forming Sodium and Potassium Hydroxide. When potassium or sodium is placed in contact with water, there is a manifest evolution of heat, for the metals are melted as they float on the surface. In this case, however, the relationship in the energy is not so easily understood as it is in the case of the formation of water, because a portion of the hydrogen is given off from the water, while potassium or sodium takes its place. To form potassium hydroxide, energy must first be added to the water in order to decompose that substance, and so expel one-half of the hydrogen. It is only after this, that potassium can take its place; but as it is manifest that heat is given off during the complete change, potassium hydroxide must have a greater heat of formation than has the water which has been decomposed by the metal. When potassium hydroxide has once been formed, it is in the same condition as the stone which has fallen to the ground; i.e., energy must be added to bring it back to its original condition. It follows that if it were practically possible to change potassium hydroxide back into water and potassium by means of hydrogen, the hydrogen, in order to effect this change, would have to be assisted by an amount of kinetic energy equal to that given off during the decomposition of an equivalent quantity of water by potassium.

If we designate a given quantity of hydrogen by X, an amount of oxygen exactly necessary to change all of this hydrogen to water by Y, and a quantity of potassium just sufficient to decompose the water so formed by Z, the compounds produced being XY (water), $\frac{X}{2}YZ$ (potassium hydroxide) then: —

I.	II.
$X + Y$	$= XY.$
Hydrogen + Oxygen	= Water.
$XY + Z$	$= \frac{X}{2}YZ + \frac{X}{2}.$
Water + Potassium	= Potassium hydroxide + Hydrogen.*

In each case the substances under I., in being converted into those under II., would give off heat; they consequently possess energy, for, when brought in contact, they are capable of performing work.

Energy Changes in the Formation of other Compounds. Substances which mutually possess chemical energy are capable of entering into chemical reactions when they are brought in contact, and the bodies produced by these reactions manifestly possess less energy than the uncombined constituents. Thus the chemical changes which take place are accompanied by an evolution of heat. It is, however, necessary in many cases to give an impulse to the substances which we intend to have enter into combination. For example, we saw that hydrogen and oxygen could be mixed without suffering any change, the chemical union of the two gases taking place only by contact with a flame or an electric spark. This condition can be compared to that of a stone at the brow of a hill, since an impulse may have to be given it before it begins to roll toward the bottom.

* In these equations the terms XY and $\frac{X}{2}YZ$ mean hydrogen and oxygen joined to form water, and oxygen, potassium, and one-half as much hydrogen united to form potassium hydroxide. The same letters separated by the sign + mean that the substances represented are mixed together, but that they have not as yet reacted to form new bodies. The quantities of matter to the left and to the right of the sign of equality are equal.

Kindling Temperatures. The great majority of bodies capable of uniting with oxygen, and of evolving light and heat during the change, must, as is the case with hydrogen, first be heated to a certain degree before the union begins to take place. The temperature at which combination with oxygen begins, is constant for a given substance, and is called the kindling temperature of that substance, while the subsequent reaction is called combustion. A little reflection will show that the kindling temperature of different bodies may vary greatly; for it is a matter of daily experience that coal has a very high kindling temperature, while phosphorus * can be ignited by the heat produced by friction.[15]

* On match-heads.

CHAPTER VII.

HYDROGEN CHLORIDE.

Common Salt. One of the chemical substances most widely distributed is common salt. It is found dissolved in sea-water, in the water of rivers, springs, and lakes, in mineral deposits often very extensive in mass, and in animal and vegetable tissues.

Preparation of Hydrogen Chloride. If salt is mixed with sulphuric acid,* there is evolved a gas with a most penetrating and acid odor. This gas is called hydrogen chloride.[16]

Properties of Hydrogen Chloride. Hydrogen chloride is a colorless gas at ordinary temperatures. It is changed to a liquid at 0° under a pressure of twenty-six atmospheres.† It is extremely soluble in water, one cubic centimetre of the latter absorbing 452 cubic centimetres of the former at ordinary temperatures. The solution has a specific gravity of 1.20, and contains 42 per cent of hydrogen chloride. "Muriatic" or "hydrochloric" acid is simply this solution more or less diluted with water. Usually it is of a yellow color, owing to impurities; but when chemically pure, it is colorless. The gas, or the concentrated solution, appears to smoke (fume) when exposed to the air, a phenomenon due to the condensation of the moisture present in the atmosphere which

* See page 56.

† The question of changing a gas to a liquid is merely one of temperature and pressure. For example, steam (gaseous water) is changed to a liquid (water) at 100°, and at the pressure of one atmosphere.

forms a solution of hydrogen chloride. When dissolving in water, hydrogen chloride causes a considerable rise in the temperature. A volume of hydrogen chloride is considerably heavier than the same volume of air, its specific gravity being 1.23 (air = 1). In order to obtain the pure gas, it must be collected by displacing dry air from a flask placed mouth upward, or by expelling mercury from a tube filled with the latter, and inverted over a trough containing the same substance. Hydrogen chloride cannot be collected over water, owing to its solubility in that liquid.[17]

The Decomposition of Hydrogen Chloride by Sodium. In studying the chemistry of hydrogen chloride, the first question to be settled is whether the gas is a compound or an element. The methods which suggest themselves are, naturally, the same as those employed for the decomposition of water. In this case, however, the sodium or potassium cannot be treated in exactly the same way as in the investigation of the chemical structure of water, because when pieces of these metals are brought in contact with dry hydrogen chloride, they soon become coated with the solid products of the reaction * which ensues, and are by this means protected from further change. We can avoid this interference by dissolving sodium or potassium in mercury,† and then substituting this solution for the pure metals. The changes brought about will not be altered by this variation in the conditions, and we have the advantage of avoiding the protective coating.

If a current of hydrogen chloride is passed through a bottle containing sodium amalgam, the gas which is col-

* In the case of water the sodium or potassium hydroxide is dissolved as fast as it is formed, and the metals are, therefore, not protected.

† The solutions produced by dissolving metals in mercury are called amalgams. The solution of sodium would thus be called sodium amalgam.

lected after the passage is no longer hydrogen chloride. It is but slightly soluble in water, it has no odor, and it burns in the air with a colorless flame — in short, it is *hydrogen*. We have already seen that this gas is not contained in the sodium; the only possible conclusion, therefore, is that hydrogen chloride is a *compound*, and that it contains hydrogen.[18]

Decomposition of Hydrochloric Acid by the Electric Current.* What are the other constituents of hydrogen chloride? The answer to this question can be found by decomposing hydrochloric acid by the electric current exactly as was done with water; and investigation has shown that a concentrated solution of hydrochloric acid in a saturated † solution of common salt is the best substance to be employed in order to study this change. If two pencils of gas carbon ‡ (connected with the poles of an electric battery) are placed in such a solution, bubbles will appear first at the negative pole, and later at the positive one.§ If two tubes filled with the liquid undergoing decomposition are inverted over the two pieces of carbon, the gases produced can be collected separately; and if the tubes are of the same size, it will be seen that *equal volumes of the two gases are collected in them during equal intervals of time.* The gas forming at the negative pole is easily shown to be *hydrogen.* That at the positive pole has a peculiar and most irri-

* The solution of hydrogen chloride in water is termed *hydrochloric acid.*

† See page 14.

‡ The hard pieces of carbon used in producing the arc electric lights.

§ The reason why the gas does not appear immediately at the positive pole is because it is to a certain extent soluble in the liquid undergoing decomposition. Only after the latter has taken up as much of this gas as it will, does the excess escape.

tating odor; it bleaches moist vegetable dyes which are brought in contact with it, and, owing to its greenish-yellow color, is called *chlorine*. Hydrogen chloride can, therefore, be decomposed into two elements, hydrogen and chlorine.[19]

Difference between the Structure of Hydrogen Chloride and of Water. There is a marked difference in one respect between the behavior of water and of hydrogen chloride toward metals. When enough sodium or potassium acts on the former to insure complete decomposition, the metal liberates but one-half of the hydrogen, while in the case of the latter it separates the entire quantity. This same distinction is shown by the relative volumes of the gases which are produced by electrolysis; for water is decomposed into *two volumes of hydrogen and one of oxygen*, while hydrogen chloride, under similar conditions, *yields equal amounts* of hydrogen and of chlorine. The hydrogen of hydrogen chloride is not, therefore, separable as two distinct and equal portions like that of water. If, then, we express the structure of water by the formula HOH (see page 31), we can express that of hydrogen chloride by H Cl.

Formation of Hydrogen Chloride from its Constituent Elements. In order to complete all of the data necessary for a knowledge of its structure, it only remains to prove that hydrogen chloride can be produced from hydrogen and chlorine. This, however, is not so easily accomplished as is the formation of water from hydrogen and oxygen, because of the practical difficulty that chlorine cannot be collected over mercury (see page 24), as it readily attacks that metal. But hydrogen and chlorine in equal volumes (produced by the decomposition of hydrogen chloride by the electric current) may be passed through a tube with a glass stopcock at each end until all of the air is expelled.* If the stopcocks are then closed, and the tube is left in diffused daylight, the hydrogen and chlorine will gradually unite, and the gas which is produced will have exactly the properties which characterize hydrogen chloride. If care is taken not to lose any of the product of the reaction, we can easily prove that the volume of gas left in the tube after the union has taken place is exactly *equal to the sum of the volumes of hydrogen and chlorine* with which it was filled in the beginning. Hydrogen chloride, therefore, is *produced* by the union of equal volumes of *hydrogen and chlorine.*[20]

The volumes of hydrogen and chlorine which combine to form hydrogen chloride are equal, no matter how the acid is produced, and, therefore, hydrochloric acid is formed by the union of definitely related masses of its constituent elements. As a given volume of hydrogen always has a constant weight when measured under the same conditions, and as this is necessarily also true of a given volume of chlorine, it follows that hydrochloric acid is produced by the interaction of definitely related weights of hydrogen and chlorine. As the volumes of these two gases when they enter into combination are equal, it must be true that the relative weights in which they combine are to each other in the same ratio as their specific gravities; that is, as 1 to 35.45 (see page 27).

Summary. From the preceding experiments we have learned the following facts: —

* *This experiment must be performed in a dark room.* A mixture of equal volumes of hydrogen and chlorine unite with a sharp explosion if they are exposed to the sunlight, or even to the rays of an arc electric light, or to those of a burning piece of magnesium.

1. Hydrogen chloride is decomposable into hydrogen *and chlorine* by means of the electric current.

2. Hydrogen chloride is produced from hydrogen and chlorine.

3. Equal volumes of hydrogen and chlorine unite to form hydrogen chloride, and the gas so formed occupies a space equal to the sum of the volumes of hydrogen and chlorine from which it is produced; or, one volume of hydrogen and one volume of chlorine form two volumes of hydrogen chloride.

4. The ratio by weight in which hydrogen and chlorine unite is as 1:35.45.

5. Hydrogen chloride possesses less chemical energy than does a mixture of hydrogen and chlorine; for if the latter mixture is brought into the sunlight, the two gases will unite with a sharp explosion, heat and light being given off at the same time.

6. Hydrogen and chlorine possess more chemical energy than does hydrogen chloride. The latter substance is extremely stable, but can be broken down by the electric current, or, in other words, by the addition of energy (see page 34).

Resemblances between Hydrogen Chloride and Water. All the above changes are exactly parallel to the similar ones which we observed while studying water; and if we measure the heat given off in the formation of a given weight of hydrogen chloride, and the heat-equivalent of the electricity necessary to decompose the same, the two will be found equal. *The heat of decomposition of hydrogen chloride is, therefore, equal to the heat of its formation* (see page 34). When hydrogen chloride is dissolved in water, considerable heat is evolved.

Therefore, the solution requires more energy for its decomposition than does the dry gas. In consequence of this fact, the solution is of necessity *more stable*. That this is the case can be shown by the following experiments: —

If a mixture of hydrogen chloride and oxygen is passed through a heated tube,* chlorine is developed, while water is formed by the union of the hydrogen of the chloride with the free oxygen which has been added. On the other hand, a solution of chlorine in water, when placed in direct sunlight, forms hydrochloric acid, and gives off oxygen.[21] In the first case, then, the hydrogen chloride and oxygen possess more energy than water and chlorine. In the second case, however, the *water* being decomposed by the *chlorine*, this relationship is exactly reversed. Water is, therefore, more stable than is dry hydrogen chloride, while, on the other hand, hydrochloric acid in solution is more stable than water. The bleaching action of moist chlorine (see page 41) can, in most cases, be referred to the effect of oxygen which is liberated from water, the oxygen being chemically more active in the instant of its separation from its compound than it is at other times.[22]

* The tube had better contain porous bodies, such as pumice stone.

CHAPTER VIII.

THE ACTION OF METALS

On Hydrochloric Acid. The Neutralization of Hydrochloric Acid by Bases.

WE have seen that when sodium acts on hydrogen chloride a change is produced similar to the one which we studied when considering the formation of sodium hydroxide from water, with the difference that, in the former case, if enough of the metal is added, the hydrogen is all separated, while *sodium chloride* remains. It is a matter of experience, however, that more metals will spontaneously react with hydrochloric acid than will decompose water.

Action of Metals on Hydrochloric Acid. Pieces of zinc or of iron,* when placed in a solution of hydrochloric acid, are instantly attacked; hydrogen is liberated, while the metals dissolve. If, after the reaction is completed, the excess of acid is evaporated, there will remain behind solid, salt-like bodies, which are chlorides of zinc and of iron respectively.[23] The parallelism between these changes and those taking place between sodium or potassium and water becomes clear if we use formulæ similar to those already employed

* The metals are practically not acted upon by water at ordinary temperatures. The solution of hydrogen chloride in water is hydrochloric acid.

(on page 31). Representing by H the one part * of hydrogen which is combined with the 35.45 parts of chlorine † in hydrochloric acid, and the corresponding chlorine by Cl, we have —

$$\mathrm{H} + \mathrm{Cl} = \mathrm{H\,Cl}.$$

Hydrogen + Chlorine = Hydrochloric acid, and

$$\mathrm{H\,Cl} + \mathrm{K} = \mathrm{K\,Cl} + \mathrm{H}.$$

Hydrochloric acid + Potassium = Potassium chloride + Hydrogen. Just as

$$\mathrm{HOH} + \mathrm{K} = \mathrm{KOH} + \mathrm{H}.$$

Water + Potassium = Potassium hydroxide + Hydrogen.

If we use zinc in place of potassium, representing by *zn* the quantity of zinc which will decompose the above weight of hydrochloric acid (i.e., which contains one part of hydrogen combined with 35.45 parts of chlorine), then ‡ —

$$zn + \mathrm{H\,Cl} = zn\,\mathrm{Cl} + \mathrm{H}.$$

Zinc + Hydrochloric acid = Zinc chloride + Hydrogen,

or, in general, using the term Me to represent any metal which is acted on by hydrochloric acid —

$$\mathrm{Me} + \mathrm{H\,Cl} = \mathrm{Me\,Cl} + \mathrm{H}.$$

In reacting with hydrochloric acid, the metals, while liberating hydrogen, form the corresponding chlorides. The chlorides can, therefore, be looked upon as hydro-

* In this book wherever proportions are indicated they are always understood to be *by weight*, unless *volume* is expressly stated.

† Hydrogen, as we have seen, combines with chlorine in the ratio of 1 : 35.45. If we select 35.45 as the weight of chlorine in the above examples, we have numbers which may be called combining weights, for they represent the ratio by weight in which hydrogen and chlorine unite. It is obviously of advantage, in this and other cases, to select the weight numbers which are determined for us by nature.

‡ The symbol *zn* is used to designate the amount of zinc which will unite with as much chlorine as will the quantity of potassium represented by the symbol K.

chloric acid, in which the hydrogen has been replaced by other metals.

Changes of Energy taking place during the Decomposition of Hydrochloric Acid by Metals. Heat is developed during the decomposition of hydrochloric acid by metals so that the energy changes during these reactions are parallel to those encountered when sodium and potassium are brought in contact with water. The heat *given off* in the formation of the metallic chlorides must, therefore, exceed that required for the decomposition of hydrochloric acid. Were it possible to convert the chloride back to the metal and hydrochloric acid by means of hydrogen, exactly as much energy would have to be added to assist this reaction as was given off in the formation of the chloride from the metal and hydrochloric acid.

In the change which is represented by the equation —

$$HCl + Me = MeCl + H,$$

the substances to the right of the sign of equality possess less energy than do those to the left. They are in the state of a stone which has fallen to the ground, for energy must be added to bring them back to their original condition. The reactions mentioned above take place spontaneously, because chemical energy can be transformed into kinetic (see page 35).

Action of Hydrochloric Acid on the Oxides of the Metals. The chlorides of the metals can be formed in another and much more general way than by the direct substitution of hydrogen: —

EXAMPLE: — If the oxide of potassium is brought in contact with hydrochloric acid, an energetic reaction at once results, much heat is developed, and, when the excess of liquid is evaporated, *potassium chloride will remain. The oxide of potassium with hydrochloric acid produces potassium chloride and water.*

If we use the same formula for oxide of potassium that we used on page 31, and the one for hydrochloric acid which we employed on page 46, then the change can be represented as follows: —

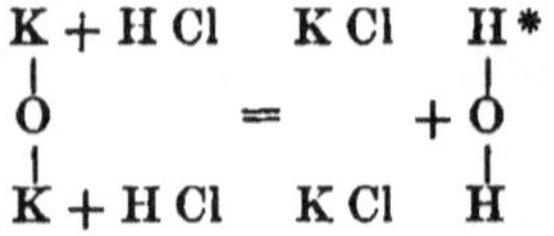

$$\begin{array}{lcccl} K + H\,Cl & & K\,Cl & & H^{*} \\ | & & & & \;| \\ O & & = & & +\,O \\ | & & & & \;| \\ K + H\,Cl & & K\,Cl & & H \end{array}$$

What is true of potassium oxide is true of most of the oxides of other metals as well, the chlorides being formed by the action of these oxides on hydrochloric acid.[24]

Action of the Hydroxides of the Metals on Hydrochloric Acid. If we compare potassium hydroxide with potassium oxide, we shall see that the former can be looked upon as water in which all of the hydrogen has been replaced by potassium, while the latter is water in which but one-half of the hydrogen has suffered the same substitution (see page 31). It seems reasonable to suppose, therefore, that potassium hydroxide, being of a nature similar to potassium oxide, will react similarly with hydrochloric acid, and, indeed, this is the fact.

Potassium hydroxide with hydrochloric acid also produces potassium chloride and water. This change becomes apparent if

* In order to secure additional clearness, the formula for potassium oxide is written in a vertical line. This obviously makes no difference. It thus becomes plain that the hydrogen of the hydrochloric acid unites with the oxygen of potassium oxide to form water. The quantity of hydrochloric acid represented by H Cl is supposed to be exactly sufficient to form potassium chloride with one-half of the potassium represented by KOK. We must, therefore, use twice this quantity to express the complete reaction. It is also clear that in our equation the hydrogen of the water formed is divisible into two halves (see page 31).

we employ formulæ similar to those already used in previous cases: —

$$KOH + H\,Cl = K\,Cl + HOH.*$$

What is true of potassium hydroxide is also true of most hydroxides of other metals, so that these hydroxides, when brought in contact with hydrochloric acid, produce the corresponding chlorides and water. The oxides and hydroxides of metals which react in the above manner with hydrochloric acid are termed *bases*.[25]

Neutralization of Hydrochloric Acid by Bases. The changes outlined above always take place between definitely related quantities of the bases and of hydrochloric acid.

This fact can readily be made apparent by experiment, but we must first procure some means of discovering with certainty the presence of a base when it is dissolved in water. This means is furnished by the use of methyl orange, a substance prepared in recent years by a complicated chemical process from one of the constituents of coal-tar. A solution of methyl orange in water is of an orange color when a soluble base is present, but turns to deep red on contact with an acid.† [26]

If a weighed quantity of potassium hydroxide is dissolved in water, a very little methyl orange added, and then hydrochloric acid is carefully run into the solution drop by drop, a point will finally be reached at which

* The quantity of hydrochloric acid here selected and represented by H Cl is supposed to be exactly enough to form potassium chloride with the potassium present in the hydroxide used. Experience teaches us that the amount of hydrogen in the acid will be exactly enough to form water with the oxygen and hydrogen present in potassium hydroxide.

† Substances which turn the red solution of methyl orange to orange yellow are called alkaline. The orange solution is said to be of an "alkaline reaction," and the substance (in this case methyl orange) which shows that the alkali is present is termed an "indicator." Substances which have the opposite effect are said to be of an "acid reaction."

the orange yellow of the "indicator" changes to red. A trace of potassium hydroxide would now restore the original color, which, again, a trace of acid would alter to red. At this point the liquid is said to be neutral. If the excess of water is evaporated from this solution, nothing but potassium chloride remains; for all the caustic potash and all the hydrochloric acid have reacted with each other to form the salt.

We have supposed that we started with a definite quantity of potassium hydroxide. Let us suppose also that the solution of hydrochloric acid is so prepared that we know exactly the amount of that substance in a cubic centimetre of liquid. It is then obvious that, by measuring the number of cubic centimetres which it takes to neutralize the base, we can ascertain the total amount of acid which is used for such neutralization. Repeated experiments will show that, with a fixed quantity of potassium hydroxide, the amount of acid necessary for neutralization is also fixed, no matter what circumstances may surround the experiments. What is true of potassium hydroxide is true of all other bases, so that—

A fixed quantity of hydrochloric acid is always neutralized by an unvarying quantity of a given base, but the amounts which will neutralize this quantity are never alike with different bases.[27]

For Example:—One part of hydrochloric acid is always exactly neutralized by

1.535 parts of potassium hydroxide,
1.096 parts of sodium hydroxide,
1.014 parts of calcium hydroxide,
0.791 parts of magnesium hydroxide,
1.360 parts of zinc hydroxide, etc.

We shall subsequently see that what is true of hydrochloric acid is true of all other acids as well; i.e., the salts are formed by the interaction of quantities of acids and bases which bear a definite relationship to each other by weight. These facts are not unexpected if we consider that every one of the chemical compounds heretofore encountered is formed by the union of *definitely related* masses of the elements.

Energy Changes during the Neutralization of Bases by Hydrochloric Acid. During the neutralization of bases by hydrochloric acid, heat is given off, so that the chloride and the water which are formed possess less energy than do the hydroxide and the acid. The substances to the right of the sign of equality in the following equation, therefore (as in the previous cases which have been cited), possess less energy than those to the left; for work must be done (energy added) to bring them back to their original condition: —

$$\text{Me OH} + \text{H Cl} = \text{Me Cl} + \text{HOH}.$$

Base (metallic hydroxide) + Hydrochloric acid = Metallic chloride + Water.

Summary. The facts which have been brought forward in the last chapter can be summed up as follows: —

1. Hydrochloric acid attacks many metals, hydrogen being liberated, and the chloride remaining.

2. Heat is given off during the changes, consequently the chlorides possess less energy than the metals and the acid.

3. Hydrochloric acid unites with the oxides of metals (bases) to produce the chlorides and water. This change is termed the *neutralization* of the base by the acid.

4. Hydrochloric acid reacts with the hydroxides of

metals (bases) to produce the chlorides and water. This is also termed *neutralization.*

5. The replacement (substitution) of the hydrogen in the acid takes place in such a way that a fixed quantity of a given metal always takes the place of a fixed quantity of hydrogen. There is a definite relationship, therefore, between the mass of the metal dissolved, the mass of hydrogen evolved, and the mass of chlorine found united to the metal in the chloride. This relationship is constant for any one metal, but is different for any two metals.

6. The neutralization of the acid always takes place in such a way that there is a definite ratio between the mass of the base neutralized and the mass of the acid used for neutralization. The chloride produced also contains the same proportion of metal to chlorine as it would contain were it formed by dissolving the metal in the acid.

7. Heat is given off during the processes of neutralization, consequently the chloride and the water which are formed possess less energy than the base and the acid before neutralization.

CHAPTER IX.

THE OXIDES OF SULPHUR.

The Chief Oxides of Sulphur, and the Acids derived from these Oxides. The Laws of Definite and Multiple Proportions. The Neutralization of Sulphuric Acid by Bases.

Acids and Salts in General. The processes of neutralization with which we became familiar during the study of hydrochloric acid are of fundamental importance in chemistry, because there is a vast number of other acids which, like hydrochloric, react with bases. Hydrogen chloride contains only two elements, but most acids are formed from three or even more. In any event, one of these elements must be hydrogen, which is replaceable by other metals. The term salt is applied to substances produced either by direct substitution of the acid hydrogen by metals, or by neutralization of bases; and a given salt has the same composition, no matter what the manner of its formation may be.

Undoubtedly the most important of acids is sulphuric acid; and, in order to understand the nature of this substance, it is necessary to inquire into the manner of its formation. This we can most successfully do by first studying the product formed by burning sulphur in oxygen.

Formation of Sulphur Dioxide. If sulphur is ignited, it burns with a blue flame, the sulphur disappears, and a gas takes its place. If the combustion is performed in a jar of pure oxygen with a sufficient quantity of sulphur, nothing but this gas (sulphur dioxide) will remain, and the *volume of sulphur dioxide produced is the same as the volume of oxygen entering into its formation.*[28]

Formation of Sulphur Trioxide from Sulphur Dioxide. A large amount of heat is given off when sulphur burns, so that the sulphur dioxide which is produced by this process obviously possesses much less energy than do the oxygen and sulphur from which it is formed. Sulphur dioxide is, however, capable of still further union with oxygen if placed under proper conditions, and the product of this union is a solid called *sulphur trioxide.*[29]

The quantities of oxygen which are united with a fixed amount of sulphur in the above two oxides are to each other in the ratio of 2 to 3; for in sulphur dioxide one part by weight of sulphur is united to one of oxygen, while in sulphur trioxide one part of sulphur is combined with one and one-half parts of oxygen.*

These facts can be summed up as follows: —

1. Each of the oxides of sulphur is produced by the union of *definitely related weights* of sulphur and of oxygen.

2. The weight of oxygen which is united to a fixed amount of sulphur in sulphur dioxide is *in simple ratio* to the weight of oxygen which is united to the same fixed quantity of sulphur in sulphur trioxide.

The statement made in the first of the above clauses is equally true, as we have seen, if it is applied to any one of the chemical compounds which we have studied;

* Sulphur dioxide, when wanted for laboratory use, is not prepared by the burning of sulphur. The method which is employed, and the means of converting sulphur dioxide into sulphur trioxide, are given in the Laboratory Appendix, experiments 28 and 29.

that in the second will be found to apply to every series of compounds in which each member is produced by the union of the same two elements. Facts so general in their application as these can be summed up in the form of *laws*, so that as the basis of chemical science we have the result of extended investigation in the following: —

1. **Law of Definite Proportions.** *Every chemical compound contains its constituent elements in an unvarying ratio by weight, no matter under what circumstances the compound may be formed.*

2. **Law of Multiple Proportions.** *If two elements (A and B) unite in* more than *one proportion, the parts of B which enter into combination with a fixed quantity of A will, in the series so produced, be in simple ratio to each other.*

Examples of Law 1.

One part of hydrogen always unites with 8 parts of oxygen to produce water.

One part of hydrogen always unites with 35.5 parts of chlorine to produce hydrogen chloride.

One part of hydrogen always unites with 39 parts of potassium and 16 parts of oxygen to produce potassium hydroxide.

Examples of Law 2.

Sulphur and oxygen produce two compounds. In one of these one part of sulphur unites with one part of oxygen to produce sulphur dioxide; in the other, one part of sulphur unites with one and one-half parts of oxygen to produce sulphur trioxide. Proportions of oxygen, 2 : 3.

Iron and sulphur produce two compounds. In one of these one part of iron unites with 1.143 parts of sulphur; in the other, one part of iron unites with 2.286 parts of sulphur. Proportions of sulphur, 1 : 2.

Nitrogen and oxygen produce five compounds. In this series one part of nitrogen is united with .569, 1.138, 1.707, 2.276, and 2.845 parts of oxygen. Proportions, 1 : 2 : 3 : 4 : 5.

Properties of Sulphur Trioxide. Sulphur trioxide is a colorless liquid at ordinary temperatures. Below 15° it changes to a solid. It boils at 46°, and forms a colorless vapor having a specific gravity of 2.76. (Air = 1.)

Formation of Sulphuric Acid from Sulphur Trioxide. Sulphur trioxide dissolves in water with a hissing noise, resembling that attending the immersion of a red-hot iron; much heat is developed, and if the excess of liquid is evaporated there remains an oil-like fluid which is termed sulphuric acid.[30] Sulphuric acid is, therefore, formed by the union of water with sulphur trioxide (which is termed an *anhydride*).

The Anhydrides of Acids. The anhydrides of acids constitute a class of bodies which has many representatives. They are oxides of not-metals which on addition of water unite with that substance to *produce the corresponding acids.*

EXAMPLES: — Sulphuric anhydride (sulphuric trioxide) by the addition of water produces sulphuric acid.

Phosphoric anhydride (produced by burning phosphorus in oxygen) by the addition of water produces phosphoric acid.

Nitric anhydride (a compound of nitrogen and oxygen) by the addition of water produces nitric acid.

Properties of Sulphuric Acid. Sulphuric acid is a colorless liquid which on superficial examination appears like an oil. When pure it has a specific gravity of 1.88 (water = 1); at 0° it crystallizes in large prisms which melt at 10°.5. It boils at 338°; but before that point is reached the acid begins to decompose into *sulphur trioxide and water*, or, in other words, in a direction which is the reverse of that of its formation. This separation is complete at a temperature somewhat above the boiling-point of the acid. Sulphuric acid has a strong inclination to take up water. During the process of solution a great amount of heat is developed; the diluted acid,* therefore, possesses much less energy than does the pure, and, as a consequence, it is less readily decomposed. Concentrated sulphuric acid violently attacks the skin and mucous membrane, so that it is a poison.[81]

Action of Metals on Diluted Sulphuric Acid. It is obvious that sulphuric acid must contain hydrogen, for it is produced by the union of sulphur trioxide with water. That this hydrogen can be replaced by other metals is proven by the following experiments: —

When diluted sulphuric acid is brought in contact with pieces of zinc, a gas is developed which can easily be identified as hydrogen. If the excess of liquid is evaporated (after as much zinc has been added as the acid will dissolve), a solid crystalline substance (zinc sulphate) will remain. Similar phenomena are observed when iron or magnesium is added to dilute sulphuric acid. With the former iron sulphate is formed, while with the latter magnesium sulphate is produced. Sodium or potassium also reacts with diluted sulphuric acid with the greatest violence, hydrogen being evolved, while sodium or potassium sulphate remains after evap-

* Dilute slowly by adding sulphuric acid to an excess of water, *not by adding water to sulphuric acid;* otherwise, the heat produced may be great enough to cause the water to boil, and in this way drops of hot acid may be scattered about.

oration. Zinc sulphate, iron sulphate, magnesium sulphate, sodium sulphate, and potassium sulphate are salts; for, as we have seen, they are produced by replacing the hydrogen of sulphuric acid by other metals.[32]

Neutralization of Sulphuric Acid by Bases. The phenomena which have just been detailed are parallel to those which we observed while studying hydrochloric acid (see page 46 and following); and in order to make this parallelism more apparent, the processes attending the neutralization of sulphuric acid should next be considered.

The oxides of zinc, iron (ferrous oxide), or magnesium, are insoluble in water; but they readily dissolve in diluted sulphuric acid. If in each case enough of the oxide is added to the acid so that a portion will remain undissolved (even after standing for some time), and if then this excess is removed by filtration, sulphates, identical with those obtained by the action of diluted sulphuric acid on the corresponding metals, will be deposited as crystalline solids upon evaporating the excess of the liquid. In the same way sodium oxide will produce sodium sulphate, and potassium oxide, potassium sulphate. Sulphuric acid is, therefore, neutralized by these oxides to form the corresponding salts and water, and similar processes can also be observed upon the solution in sulphuric acid of a very large number of the oxides of other metals.[33]

Difference between the Neutralization of Sulphuric Acid and of Hydrochloric Acid. So far we have observed no distinction between the action of sulphuric acid and hydrochloric acid during the process of neutralization.

That there is a marked difference, however, will become apparent as soon as we study the neutralization of sulphuric acid by means of sodium or potassium hydroxide. We weigh off sodium hydroxide and sulphuric acid in such proportions, that for one part of the former we have 2.45 parts of the latter, and then, after dissolving the sodium hydroxide in water and diluting the sulphuric acid with the same, bring the two solutions in contact. All the sulphuric acid and all the sodium hydroxide will enter into the reaction, and a salt will be produced. This salt can be isolated by evaporating the excess of liquid. It differs from those we have previously encountered in the fact that, in spite of its being formed by the action of an acid on a base, it still contains replaceable hydrogen.* This salt, therefore, has resulted from replacing by sodium only a portion of the available hydrogen of sulphuric acid. That this is a fact can readily be shown by dissolving the salt in water, adding a second quantity of sodium hydroxide equal to the first, and again evaporating the excess of liquid. There will remain a sodium sulphate, which no longer contains replaceable hydrogen; and this substance differs radically in properties from the one which was isolated in the first instance. The salts which are produced by replacing only *one-half* the hydrogen of sulphuric acid by metals are termed *primary* or *acid;* those formed by replacing *all* the hydrogen, *secondary* or *neutral* sulphates.[34]

Comparison of the Structure of Sulphuric Acid and of Water. In its structure sulphuric acid can be compared to water; for its hydrogen is divisible into two

* I.e., hydrogen which can be replaced by other metals to form a salt.

halves by means of metals, just as the hydrogen of water is. This parallelism will be made clear if we consider the following two formulæ (X representing what is united with hydrogen to form sulphuric acid): —

$$\underset{\text{Water.}}{H - O - H} \quad \text{and} \quad \underset{\text{Sulphuric acid.}}{H - X - H}$$

The substitution of hydrogen by sodium in these two substances can be expressed as follows: —

1. $$Na^{*} + HOH = Na\,OH + H.$$
Sodium + Water = Sodium hydroxide + Hydrogen.

$$Na\,OH + Na = Na\,O\,Na + H.$$
Sodium hydroxide + Sodium = Sodium oxide + Hydrogen.

2. $$Na + HXH = Na\,XH + H.$$
Sodium + Sulphuric acid = Primary sodium sulphate + Hydrogen.

$$Na\,XH + Na = Na\,X\,Na + H.$$
Primary sodium sulphate + Sodium = Secondary sodium sulphate + Hydrogen.

Comparison of the Structure of Hydrochloric Acid with that of Sulphuric Acid. When sodium hydroxide is brought in contact with *twice* as much hydrochloric acid as it is capable of neutralizing, one-half the available hydrogen will be replaced by the sodium; but this hydrogen will be the constituent part of only one-half the hydrochloric acid, while the other half will remain unchanged. With sulphuric acid, on the other hand, if the same quantitative relationship between the base and the acid is retained,† *one-half* the hydrogen will also be replaced; but, at the same time, *all* the sulphuric acid will have

* The quantity of sodium represented by the symbol Na is, as in previous instances, supposed to be exactly enough to substitute *one part by weight of hydrogen.*

† I.e., if *one-half* as much sodium hydroxide as is necessary for complete neutralization of the sulphuric acid is added.

entered into the reaction and so have produced the primary sulphate of sodium. An acid like hydrochloric acid is termed a *monobasic* acid; an acid like sulphuric acid is termed a *dibasic* acid. What is true of hydrochloric acid and of sulphuric acid is equally true of all monobasic and diabasic acids.

The Sulphates are formed by the Interaction of Definitely Related Weights of Bases and of Sulphuric Acid. The method by which we can ascertain that an unvarying amount of a given base is always necessary to neutralize completely a fixed quantity of sulphuric acid is identical with that employed in the study of hydrochloric acid (see page 49). A little methyl orange (page 49) is added to dilute sulphuric acid, and then potassium hydroxide or sodium hydroxide solution is added until the color just changes from deep red to orange yellow. At this point the sulphuric acid will be completely neutralized, and the *secondary sulphate* will be formed. If we start with a definite quantity of the base, and if the solution of sulphuric acid is so prepared that we know exactly the amount of that substance in a cubic centimetre of the liquid, it is obvious that, by measuring the number of cubic centimetres which it takes to neutralize the base, we can ascertain the total amount of the acid used for such neutralization. If this method is carefully followed out, a series of experiments will develop the following law.[35]

A fixed quantity of sulphuric acid is always completely neutralized by an unvarying quantity of a given base, but in *no two different bases* are the amounts which will neutralize this quantity of acid alike.

Examples: — One part by weight of sulphuric acid is always completely neutralized by —

1.143 parts by weight of potassium hydroxide,
0.817 parts by weight of sodium hydroxide,
0.755 parts by weight of calcium hydroxide,
0.595 parts by weight of magnesium hydroxide,
1.013 parts by weight of zinc hydroxide, etc.

If we compare the above table with the similar one on page 50, in which the quantities of the same hydroxides necessary to neutralize one part of hydrochloric acid are given, there will be brought to light the remarkable fact that *the amounts of the hydroxides are in the same ratio.* This fact will become apparent if we multiply all of the numbers in the first table by 36.5, and all of those in the second by 49;* for then: —

56 parts of potassium hydroxide,
40 parts of sodium hydroxide,
39 parts of calcium hydroxide,
29.5 parts of magnesium hydroxide,
49.65 parts of zinc hydroxide,

are completely neutralized either by 36.5 parts of hydrochloric acid, or by 49 parts of sulphuric acid. If similar investigations are conducted with any other acid, the above parts of the hydroxides are also the quantities which are necessary to neutralize that particular amount of *any* acid which contains one part of hydrogen; and from this it follows that the reverse must also be true. We could, therefore, construct a table of acids with numbers so selected that each would repre-

* The reason for selecting 36.5 and 49 as the numbers with which to multiply the quantities is that 36.5 and 49 are the parts by weight of hydrochloric acid and of sulphuric acid which contain one part by weight of hydrogen.

sent a weight of acid containing one part of hydrogen, and then any one of those quantities would exactly neutralize any one of the amounts given in the above table of hydroxides. It is, of course, understood that all these weights are to be measured by the same unit. The relative weights of bases and acids so ascertained are termed their *equivalent* weights.

Relationship between the Amounts of Base Necessary to form the Primary and Secondary Sulphates. It has been shown that the hydrogen of sulphuric acid is, like that of water, divisible into two equal parts by the action of sodium or potassium hydroxide. We should therefore select, as a basis for consideration, not the equivalent weight of sulphuric acid, but twice that amount, or, in other words, that part by weight of sulphuric acid which contains two parts of hydrogen. If we do this, we find that 98 parts of sulphuric acid react with

56 parts of potassium hydroxide and
40 parts of sodium hydroxide

to form the *primary* sulphates of these metals, and with

112 parts of potassium hydroxide and
80 parts of sodium hydroxide

to form the *secondary* sulphates.

These quantities of potassium and of sodium hydroxide are, however, in the simple ratio of 1 : 2; so that the law of multiple proportions (page 55) applies to the more complicated processes of neutralization as well as to the comparatively simple ones of the union of the same two elements with each other to form more than one compound.

Summary.

1. Sulphur, in burning, unites with oxygen to form sulphur dioxide.

2. Sulphur dioxide can further take up oxygen, under proper conditions, to form sulphur trioxide.

3. The parts of oxygen which are united with a fixed quantity of sulphur in these oxides are to each other as 2 : 3.

4. Sulphur trioxide unites with water to produce sulphuric acid. A large amount of heat is given off during this change, and the same is true if the acid is diluted with water. Diluted sulphuric acid, consequently, is more difficult to decompose than concentrated.

5. Diluted sulphuric acid when in contact with certain metals liberates hydrogen while the corresponding sulphates are formed. The hydrogen of the acid is replaced by the metals.

6. Diluted sulphuric acid reacts with the *oxides* or *hydroxides* of metals (bases) to form sulphates and water.

7. Sulphuric acid can form two sulphates each of sodium and of potassium. These sulphates differ as follows: In one class (primary or acid sulphates), only one-half the hydrogen contained in the sulphuric acid from which the salts are formed is replaced by the metal. In the other class (secondary or neutral sulphates), all the hydrogen is replaced by the metal. Sulphuric acid is a dibasic acid.

8. Hydrochloric acid differs from sulphuric acid; for when brought in contact with sodium or potassium hydroxide it forms but one kind of chloride, i.e., the chloride produced by replacing all of the hydrogen of

the hydrochloric acid which is neutralized. If an excess of hydrochloric acid is present, this excess remains unaffected by the base. Hydrochloric acid is a monobasic acid.

9. The sulphates are formed by the interaction of definitely related amounts of base and acid.

10. The ratio between the weights of various bases, which are so related in quantity that they will exactly neutralize a given weight of one acid, is also the ratio by weight in which these same bases will exactly neutralize a fixed amount of any other acid.

CHAPTER X.

THE ATMOSPHERE.

Physical Properties of the Atmosphere. The Barometer. Boyle's and Charles's Laws.

In the previous portions of this book mention has been made of the fact that substances which burn in oxygen will do the same thing, with somewhat diminished intensity, in the air. As by far the greater number of combustions take place in the latter medium, and as it plays such an important part in the most frequently recurring chemical phenomena, it is advisable to become acquainted with the composition and properties of the atmosphere at an early period in the study of chemistry. This is the more necessary because one of its constituents (nitrogen) is an element with some of whose compounds we must soon become familiar.

The Weight of the Atmosphere. The colorless, gaseous envelope which surrounds the earth is termed its *atmosphere.* The atmosphere, being a material substance, possesses weight, and by reason of this weight it exerts a pressure on all things beneath its surface. That this is the case can readily be shown by the following experiments: —

A piece of sheet rubber is tied over one end of a glass cylinder, while the other end, ground so as to fit air-tight over the plate of an air-pump, is placed over the opening which connects with the pump. The sheet of rubber will be pressed inward as soon as part of the air in the cylinder is exhausted. Two hollow hemispheres, which fit air-tight along their edges, are held firmly together when the air within them is rarefied. The above phenomena are not observed until the air is exhausted; but this is due to the fact that the atmosphere, being perfectly elastic, presses equally in all directions. It can, therefore, make its pressure manifest only when a difference is established.

The Barometer. If a tube is entirely filled with liquid, and the lower end is then opened under the same fluid while the upper end remains closed, the liquid will not drop from the tube, but will remain in equilibrium at a certain height. This is owing to the pressure exerted by the surrounding atmosphere upon the surface of the fluid in the exterior vessel. The height of the column which can be so maintained is inversely proportional to the specific gravity of the liquid employed. It varies with the atmospheric pressure, and is therefore used to measure it. If a tube is longer than this column, then a space containing no air will remain above the liquid in it. This space is termed a *Torricellian* vacuum. An instrument called the barometer is constructed on the above plan, the fluid in use being mercury; and the distance from the upper level of the meniscus of the mercury in the tube to the upper level of that in the container underneath is termed the height of the barometer. The mean height of the barometer at the level of the sea is 760 millimetres, and this is taken as the standard for all measurements. The pressure exerted by a column of mercury 760 millimetres high on a square centimetre is 1.033 kilograms. The standard pressure of the atmosphere upon the same surface is, therefore, equal to the same quantity. If this pressure is diminished, the mercury will fall; if it is increased, it will rise. If, therefore, the barometer is taken from the sea level to a point above the same, a diminution in the height of the column corresponding to a diminution in the weight of the atmosphere above it will be observed. The barometric height at any place is frequently changing because the atmosphere is not at rest, and is subject to variations in density, owing to differences in temperature. When any portion of the atmosphere is warmed, its density diminishes; the warm air will rise, its pressure will diminish, and, as a consequence, the barometer will fall. The same result is brought about if the temperature remains constant while that of contiguous portions of the atmosphere is lessened.[86]

Relation of the Volume of a Gas to Pressure. If a given volume of air (V) is subjected to a certain pressure (P), the volume will be diminished as the pressure is increased. If the second volume produced by this increased pressure is termed V′ and the second pressure P′, then the relation —

$$V : V' :: P' : P$$

will be realized. The same is true of all other gases which, like the atmosphere, are at a temperature far above the point at which they become liquid. This fact can be summed up in the following law:[37] —

Boyle's Law. *If the temperature remains constant, the volume of a given quantity of gas varies inversely as the pressure to which it is subjected.*

EXAMPLES: —

Let P = the pressure of one atmosphere;
Let P′ = the pressure of two atmospheres,
V = 10 cubic centimetres,

then, $V' = \frac{VP}{P'} = \frac{10}{2} = 5$ cubic centimetres. If $P' = 3$ atmospheres, then $V' = 3\frac{1}{3}$ cubic centimetres, and so on.

If P is equal to the pressure of one atmosphere, it is equal to the pressure exerted by 760 mm. of the barometer. If h represents the height in millimetres of the observed barometric pressure, and V_0 the volume of a gas at 760 mm. pressure, and V_t that volume at the observed pressure, then, $V_0 : V_t :: h : 760$, from which it follows that—

$$V_t = \frac{V_0 \times 760}{h} \quad \text{(Equation 1.)}$$

Application of Boyle's Law in the Calculation of Gas Volumes. To find the volume which any gas occupies at other than the standard barometric pressure, we must multiply its volume at the standard pressure by 760, and divide by the observed height of the barometer.

From equation 1 it follows that—

$$V_0 = \frac{V_t\, h}{760} \quad \text{(Equation 2.)}$$

To find the volume of any gas at standard barometric pressure, we must multiply the observed volume by the observed barometric pressure in millimetres, and divide by 760.

Law of Charles and Gay Lussac. *All gases when not too near the point at which they liquefy, expand very nearly* $\frac{1}{273}$ * *of their volume for each increase of* 1° *centigrade in temperature.* This fraction is called the co-efficient of expansion of gases, and is independent of the pressure to which the gas is subjected.

Examples:—A litre of gas at 0° will be $1 + \frac{1}{273}$ at 1°, $1 + \frac{2}{273}$ at 2°, ten litres would be $10 + \frac{10}{273}$ at 1°, and so on, so that—

$$3.\ V_t = V_0 + V_0\, \alpha\, t$$
$$4.\ V_t = V_0 (1 + \alpha\, t)$$
$$5.\ V_0 = \frac{V_t}{1 + \alpha\, t}$$

where V_0 = the volume of any gas at 0°, V_t the volume of the gas at the temperature t, and $\alpha = \frac{1}{273}$.

* In exact numbers .00367.

To calculate the volume which a gas would occupy at 0° (the standard temperature), we must divide the observed volume by $1 + \alpha t$.

When a gas is at ordinary temperature and pressure, and we wish to ascertain its volume at standard barometric pressure (760 millimetres) and temperature 0°, corrections must be made both for temperature and pressure. This can be done in one operation by combining equations 2 and 5, which gives —

$$6.\quad V_0 = \frac{V_t\, h}{760\,(1 + \alpha t)}$$

If a gas is introduced into the Torricellian vacuum so that a part of the mercury is expelled from the tube, this gas is obviously under atmospheric pressure, minus the height of the mercury still in the tube. If this latter quantity is w, then equation 6 becomes —

$$7.\quad V_0 = \frac{V_t\,(h - w)}{760\,(1 + \alpha t)}$$

Standard Conditions for the Measurement of Gas Volumes. A gas at temperature 0° and pressure 760 millimetres is said to be under standard conditions; and all comparisons between the volumes of different gases presuppose that these volumes are under those conditions, unless the contrary is expressly stated.

Vapor Pressure and Correction for the same. Gases which are undergoing observation may contain water. If this is the case, a further correction must be made for the pressure of the vapor of water, since it is obvious that this vapor exerts a pressure which increases with the temperature exactly as does that of the gas which is to be measured. This amount, in millimetres, must therefore be deducted from the observed barometric height, so that equation 7 becomes —

$$8.\quad V_0 = \frac{V_t\,(h - w - \phi)}{760\,(1 + \alpha t)}$$

where ϕ represents the vapor pressure in millimetres at the temperature of observation. What is true of water vapor is true of the vapors of other liquids as well, with the distinction that at the same temperature the vapors of different liquids exert different pressures.* As a matter of fact, water vapor and, at high temperatures, the vapor of mercury, are the only ones to be considered in elementary work.[86]

* A table giving the tension of water vapor at different temperatures can always be found in the larger manuals of chemistry, or in small volumes of "tables" for chemists. One of the latter should always be easily accessible.

CHAPTER XI.

THE ATMOSPHERE (*Continued*).

Combustion in Oxygen and in Air.

Isolation of Nitrogen from the Atmosphere. Heat to redness a long tube containing a quantity of fine copper shavings. Pass air through this tube, and collect the gas that emerges in a tube closed at one end, filled with water and inverted in a water trough. This gas will be found to differ in properties from the air which originally contained it. It is colorless and odorless. A burning pine splinter placed in it is instantly extinguished. At the same time, unlike hydrogen, it does not itself burn. This gas is an element and is termed nitrogen.

Properties of Nitrogen. Nitrogen has a specific gravity (air ‖ 1) of .97209. At a pressure of 35 atmospheres and at temperature of − 146°.3 it becomes liquid. This liquid boils at − 193° and solidifies at − 203°. Chemically, nitrogen in a free state is a remarkably indifferent substance, uniting with other elements only under the greatest provocation. Owing to this indifference, it will neither burn nor support combustion, and animals are suffocated when placed in it.

The Atmosphere contains Nitrogen and Oxygen. The hot copper over which the air was passed is completely changed, and at the same time it has gained in weight. The same alteration in the copper can be brought about by heating it in pure oxygen. Hence, we conclude that in the above experiment it has been oxidized by the air to form copper oxide. The atmosphere, therefore, contains nitrogen and oxygen.[39]

Combustion in Oxygen. Oxygen is a supporter of combustion. This means that many substances, if they are heated in the gas to points above their kindling temperatures, will unite with it energetically, evolving heat and light (see page 36).[40]

Examples: — A small piece of phosphorus, if ignited in the air and placed in a jar of oxygen, will burn with dazzling brilliancy, the product of the combustion being an *oxide* of phosphorus (phosphorus pentoxide).

Sulphur, ignited in the air, will burn energetically in oxygen, producing an oxide of sulphur (sulphur dioxide; see page 53). A piece of charcoal heated to redness will burn brightly in oxygen, the resulting compound being an oxide of carbon (carbon dioxide).

In the same way many other elements, sodium, potassium, magnesium, will burn in oxygen with the greatest energy, while in each case the corresponding *oxide* is produced; and even substances which under ordinary circumstances are considered as being non-combustible (for example, a steel watch-spring) can be ignited in oxygen.

Changes of Energy during Combustion. The cause underlying all these phenomena is the same, and is identical with the one discussed when we studied the formation of water during the combustion of hydrogen in oxygen. The substances which burn all possess *chemical energy* when in the presence of oxygen. This chemical energy is converted into kinetic energy (light and heat) during the combustion, so that the resulting bodies possess less energy than do the substances from which they are produced. Obviously, therefore, as mucn energy would have to be added to decompose these substances as is given off in their formation, provided equal quantities of matter are considered in each case (see page 34).

Slow Oxidation. Oxidation may take place so slowly that the heat given off during any given time is too small to be observed by the senses, and can be measured only by apparatus specially constructed for the purpose. The *total amount* of kinetic energy manifested during such a change is, however, the same as it would be were the same amount of the same substance burned rapidly, provided that the oxides produced are identical. The amount of energy employed to decompose a given quantity of these oxides is unvarying, no matter whether they have been produced rapidly or slowly. A parallel condition is found in that of a weight which has been lowered from a height. It obviously requires a certain expenditure of energy to bring this weight back to its original position, and this expenditure is entirely independent of the circumstances attending the fall. We have already seen that the heat given off during the *formation* of a chemical compound is equivalent to the amount of energy necessary for its *decomposition* (see page 34).

Incandescence during Combustion. Production of a Flame. The phenomena of combustion are, therefore, caused by the union of certain substances with oxygen, the change taking place in so short an interval of time that the heat caused by the reaction is not conducted away with sufficient rapidity, and raises the temperature to a point at which the burning body becomes incandescent.

When the ignited substance is a gas, or when in burning it either gives off a gas or is converted into a gas, a flame is produced, as in the case of the burning of phosphorus or sulphur. On the other hand, when no vapor is present, the burning substance simply glows, as is seen during the combustion of carbon. A flame is, therefore, developed by an incandescent gas.

The Phenomena of Gaseous Combustion are Reversible. Where a gas is capable of uniting with oxygen with sufficient energy to produce a flame, it is a matter of indifference which of the two gases forms the entering, and which the surrounding medium, since the phenomena of combustion are caused solely by the chemical union of the two. The terms "combustible" and "a supporter of combustion," as applied to gases, are used simply because it is more usual to see gases burning in oxygen or air than it is to see oxygen or air burning in other gases.[41]

EXAMPLES: — Hydrogen burns quite readily in oxygen with a very hot flame, and, *vice versa*, oxygen will burn in hydrogen with an equally hot flame. In both instances water is produced. Sulphur will burn in oxygen, and, on the other hand, oxygen will burn in the vapor of sulphur, the same oxide of sulphur being formed in each case.

Union with Oxygen is not Necessary to cause Incandescence. Any two elements which are capable of directly uniting, give off heat during such union; and this heat may be sufficient to cause both substances to become incandescent.

EXAMPLES: — Hydrogen, when heated to its kindling temperature, will readily burn in an atmosphere of chlorine, the product being hydrogen chloride; and, of course, chlorine will also burn in hydrogen.

Phosphorus is capable of burning in chlorine to form a chloride of phosphorus, just as it will burn in oxygen to form an oxide.

Iron filings, when mixed with finely divided sulphur, will unite with the latter with the greatest energy, if sufficient heat is applied to melt the sulphur, and thereby cause an intimate contact between the two elements. The mixture becomes brightly incandescent, and sulphide of iron is formed as a product of the reaction.

General Nature of the Phenomena of Combustion. The phenomena of combustion in oxygen are therefore merely special cases of chemical reactions which are very general in their character. The fact that enough heat is given off during these changes to cause the reacting elements to glow is incidental, and not essential to the character of the phenomena as a whole. The essential feature is that elements which are capable of direct union possess chemical energy, and this chemical energy is converted into kinetic energy when such union takes place. We have two conditions in which the elements can be encountered: —

1. Those in which they are chemically separate;
2. Those in which they are chemically combined;

and if the union of the elements can be brought about without the addition of external energy, then those substances under the second head possess less chemical energy than do those under the first.

Enduring Nature of Chemical Energy. Chemical energy is a most enduring form of energy. A piece of coal or of sulphur may remain buried in the earth for years without the least loss of the power which these substances are capable of yielding when burned. Chemical energy is also the most concentrated form of energy. The amount of heat given off by the combustion of one gram of hydrogen would be capable of raising 14,000 kilograms through the distance of one metre if this heat could be completely converted into mechanical work. At the present time no other form of energy combines so great convenience and permanence in transportation as does chemical. Steamboats crossing the ocean take along their stock of chemical energy in the form of coal, which, by the heat given off in its combustion, furnishes the power to propel the vessel. In former geologic periods the flourishing vegetation stored up this force by making use of the radiating energy of the sun; this vegetation was destroyed, buried by the *débris* of succeeding catastrophes, and is now, in the form of coal, finding its uses in the daily avocations of men.*

Chemical Energy is also possessed by Chemical Compounds. The separate elements which are capable of chemical union are not the only forms of matter which possess chemical energy. Many chemical compounds also unite, with the evolution of heat and even of light, to produce new compounds possessing less chemical energy than did those from which they were formed. We have studied examples of such changes in the neutralization of acids by bases, and in the union of salts with their water of crystallization. In short, the tendency of substances which possess chemical energy is so to interact with each other as to transform the whole or a part of that chemical energy into kinetic energy, or, in other

* This paragraph is in part taken from Ostwald, *Lehrbuch der Allgemeinen Chemie*, 2[1], 53.

words, to pass from a state of higher energy to one of a lower. Those changes attendant on combustion of oxygen are, therefore, only specific cases of what in reality is most general in chemistry. They have assumed such great relative importance only because of the general distribution of the element oxygen, and because of its essential bearing upon life. From a scientific point of view, any one of the parallel chemical reactions in which the union of elements is attended with the evolution of light and of heat is as important in illustrating a general law as is the union of elements with oxygen.[43]

Combustion in Air. Substances which burn in oxygen will also burn in the air, but with diminished intensity, owing to the fact that the oxygen is diluted by the indifferent gas, nitrogen. If the combustion takes place in a closed space, it will continue until all, or nearly all, the oxygen is used up, and then, provided the products of combustion are solid, nothing but nitrogen will remain.

Examples: — A piece of phosphorus ignited and placed in a closed jar of air will continue to burn until the oxygen is exhausted. The oxide of phosphorus produced in this way is identical with the one formed in oxygen. The gas which remains has all of the properties of nitrogen; it will neither burn nor support combustion.

A burning candle placed in a closed air-space will gradually be extinguished after using up the available oxygen. The gases which remain consist of nitrogen, carbon dioxide, and some oxygen.*

* All the oxygen is not used by the burning candle, or by an animal placed in a closed air-space. The extinction of the flame and asphyxiation take place while a considerable quantity of oxygen is still uncombined.

This mixture will neither burn nor support the combustion of a candle.[44]

An animal placed in a closed air-space will become unconscious and finally die. The oxygen inhaled by the lungs is in part absorbed by the blood, and is used in the body for purposes of oxidization. As the tissues of the body consist mostly of compounds of the element carbon with hydrogen, nitrogen, sulphur, etc., the products of this oxidization are in large part carbon dioxide and water vapor. The air exhaled from the lungs contains less oxygen and more carbon dioxide than does that which is inhaled. All the oxygen taken in, however, is not consumed in the production of these two gases; for there are other constituents of the body (sulphur, phosphorus) which are also oxidized and separated by different channels, while a portion of the oxygen is returned to the air unchanged. As the oxygen of the air becomes used up by a breathing animal, and is in part replaced by carbon dioxide, the closed air-space becomes unfit for the support of animal life long before all of the oxygen is consumed. This takes place the more rapidly because, in addition to the substances mentioned, there are other gases of a poisonous nature exhaled from the lungs.

The Process of Breathing is Analogous to Combustion. The process of breathing can, therefore, be compared to combustion. The animal and the surrounding oxygen possess chemical energy which is converted into kinetic energy by respiration and its consequences. This kinetic energy is manifested as heat, so that the body preserves a temperature above that of the surrounding atmosphere, while the substances produced possess much less chemical energy than did those which were originally present.

The Total Heat given off in Combustion in Air is the same as that given off in Combustion in Oxygen. A number of other examples of combustion in air could be mentioned; but, as they do not differ in principle from combustion in oxygen, it is not necessary to enumerate them. The total amount of heat given off during the combustion of a given weight of a substance is the same, whether it is burned in air or in oxygen. This fact is obvious when we consider that it will take exactly as much energy to decompose the body which is formed as has been given off in its production. The amount necessary for this decomposition is manifestly independent of the circumstances surrounding the formation.

CHAPTER XII.

THE ATMOSPHERE (*Continued*).

Composition of the Atmosphere.

WE have learned that the atmosphere contains oxygen and nitrogen, but we have not ascertained the relative proportion of the two gases, nor have we decided whether they are chemically combined or present merely as a mechanical mixture.

Estimation of the Relative Amounts of Oxygen and Nitrogen in the Atmosphère. A rough means of estimating the amount of oxygen and nitrogen in the atmosphere is found in the following experiment: —

A glass tube sealed at one end is divided into five equal parts by means of five rubber rings slipped over the outside. This is inverted over a cylinder of water so that the level of the liquid within and without is at the first ring, and the tube is then firmly clamped in this position. A piece of phosphorus is fixed on the end of a long, sharp-pointed wire, and then the wire is bent so that the phosphorus can be thrust up into the tube. Slow oxidization of the phosphorus will set in, and the oxide of phosphorus produced will be dissolved by the water, which will rise in the tube so as to take the place of the oxygen absorbed. The reaction is complete after some days; and then it will be discovered, after the tube has been lowered so that the level of the liquid without and within is again the same, that four-fifths of the air have remained unabsorbed. Approximately, therefore, by bulk one-fifth of the atmosphere consists of oxygen, and four-fifths of nitrogen.[45]

More accurate results can be obtained by the use of the eudiometer tube (described on page 24, and in Experiment 10 of the Appendix). This is filled with so much mercury, that, when it is inverted over the mercury trough, about fifteen cubic centimetres of air will remain enclosed. This volume is recalculated to standard conditions by noting the temperature, barometric pressure, and height of the column of mercury above the trough (see page 68 and following). About seven cubic centimetres of pure hydrogen are now added; and the volume of this gas is also, after making the necessary observations, recalculated to standard conditions. The mixture is now exploded by means of an electric spark. The contraction in volume which results is due to the formation of water from the union of hydrogen and oxygen; and as two volumes of hydrogen combine with one of oxygen to produce water, it follows that one-third of the total diminution in volume must represent the amount of oxygen which was originally present.[46]

Relative Volume of Oxygen and Nitrogen in the Atmosphere. Careful investigations by this means have shown that the volume of oxygen in ordinary out-door air varies from 20.3 to 21.5 volumes in 100.* This variation is sufficient to show us that in all probability the atmosphere is not a chemical compound; for, as we have seen, true chemical compounds are characterized by an unchanging composition (see page 55). This probability is transformed to a certainty by the following experiment: —

Proofs that the Atmosphere is not a Chemical Compound. A given volume of water will dissolve quite a little more oxygen than nitrogen, so that if we collect the

* I.e., in 100 cubic centimetres of air there are 20.3 to 21.5 cubic centimetres of oxygen.

gas which passes off from water which has been exposed to the air under pressure, it will contain a greater proportion of oxygen than it did before. Using the air which has thus been altered in composition, this operation can be repeated with a similar result until nearly pure oxygen is finally obtained.

As a final proof it may be mentioned that oxygen and nitrogen, mixed in the proper proportions, form air without alteration either in the total volume of the two gases or of the temperature, provided they were under like conditions before being brought in contact. The atmosphere is, therefore, a mechanical mixture of oxygen and nitrogen, containing in round numbers one part by volume of the former to four parts of the latter.

Other Substances Present in the Atmosphere. In addition to the two fundamental gases, other substances are present in the air in small quantities. These may be of two kinds, — gaseous and solid. The chief gaseous impurities are water vapor, carbon dioxide (see page 154), and compounds of ammonium (see page 99). The solid substances are present as dust.

Water Vapor in the Atmosphere. The quantity of water vapor present depends upon the temperature, season of the year, and locality; but it is seldom of sufficient amount to cause the atmosphere to be completely saturated. The evaporation of oceans, lakes, and rivers, furnishes a never-ending supply of water, which is increased by hot weather, and is generally greater by day than by night. If the atmosphere is nearly saturated with moisture, any diminution in the temperature will cause a fall of rain or the formation of dew. Very little evaporation can take place while such a condition prevails. Drops of water collect upon a cold surface because the air in the immediate neighborhood is cooled below the point at which it is saturated with vapor. The temperature at which these drops collect is called the dew-point; and as the maximum amount of water vapor which can be contained in a given volume of the atmosphere at any definite temperature is known, the discovery of the dew-point affords a ready means of ascertaining the amount of moisture really present in the air. The ratio between the tension of water vapor (see page 70), which would be found were the air fully saturated at the prevailing temperature, and that tension which really exists, is called the relative humidity.[47]

Water in the atmosphere is absolutely essential to plant life. The liquid falls upon the soil as rain, and is absorbed by the radicles of plants. Afterwards it circulates through the entire system of the plant, taking part in various physiological changes, and finally evaporating from the leaves. The amount of moisture which passes from large areas covered by vegetation is enormous, so that wooded districts cause an equitable distribution of rain.

Carbon Dioxide in the Atmosphere. Carbon dioxide is invariably found in the air, and is as important to living organisms as oxygen itself. It is constantly added to the atmosphere by burning fuel, from volcanic craters and fissures, from the breathing of animals, and from decomposing organic matter. If no means were provided for the removal of carbon dioxide from the atmosphere, the increase would soon destroy all living organisms dependent upon respiration. Fortunately plants growing in the sunlight absorb carbon dioxide from the air, using for this absorption a green coloring-matter which is contained in the leaves, and which can eliminate oxygen from, and add hydrogen to, carbon dioxide. By this means a substance is produced which can form all the innumerable compounds of carbon, hydrogen, and oxygen occurring in the vegetable kingdom.[48]

The Quantity of Carbon Dioxide in the Atmosphere. The quantity of carbon dioxide in a given volume of air varies slightly; but, normally, it is about three parts in ten thousand. It seems that the proportion of carbon dioxide is greater at night than in the daytime, and in summer than in winter. The amount of carbon dioxide in crowded rooms is increased by the breathing of people within the closed air-space, yet this does not often take place to such an extent that the oppressive feeling caused by such an atmosphere can be ascribed entirely to this increase. The unpleasant effect is due in part to exhalations of organic matter which passes off from the lungs.

The presence of carbon dioxide in the atmosphere can be proved at any time by exposing some clear lime-water to the action of the air. If carbon dioxide is present, a white crust of the carbonate of calcium will be found upon the surface in a short time.

Ammonia in the Atmosphere. Another impurity present in the atmosphere is ammonia, which is found combined with acids, as ammonium carbonate, nitrate, or nitrite. These substances are washed into the soil by rains, and are then taken up by plants to form essential portions of their tissues which contain nitrogen. The amount of ammonium compounds in the atmosphere is very small, not more than forty parts by weight in one million parts of air.

The solid particles of dust floating in the air may be such substances as common salt (sodium chloride), or they may consist of micro-organisms, which not infrequently are capable of inaugurating disease.

Summary of the Chapters on the Atmosphere.

1. The atmosphere is composed of oxygen and nitrogen, with small quantities of other substances. (Carbon dioxide, water vapor, ammonium salts, solid particles.)

2. Nitrogen is chemically an indifferent gas, and does not support combustion.

3. Oxygen is a supporter of combustion; i.e., many substances will burn in the gas, producing oxides.

4. Substances which burn possess chemical energy when in the presence of oxygen. This chemical energy is converted into kinetic energy during the combustion.

5. The substances produced by the union possess less chemical energy than do those from which they are formed.

6. Oxidation may be slow or rapid. The amount of chemical energy converted into kinetic energy is, however, the same in either case, provided the amounts of the substances used are the same and the products of combustion are identical.

7. In rapid combustion the temperature is raised to a point at which the reacting body becomes incandescent. If the burning substance is a gas, or gives off a gas, then a flame is produced. It is a matter of indifference which gas is entering, and which forms the surrounding medium, the combustion taking place just the same in either case.

8. The phenomena of combustion in oxygen are only special cases of chemical reactions which are general in their character; similar changes, therefore, take place when bodies unite with substances other than oxygen.

9. We have two conditions in which the elements are encountered, — those in which they are chemically separate, and those in which they are chemically combined. The separate elements are not the only forms of matter which possess chemical energy. Chemical compounds can also unite, with the evolution of light and heat.

10. Substances which burn in oxygen will also burn in air with diminished brilliancy. The process of breathing can be compared to combustion.

11. The total amount of heat given off during the

combustion of a given weight of a body is the same whether it is burned in air or in oxygen.

12. The oxygen in the atmosphere varies from 20.3 to 21.5 volumes in 100. The air is a mixture of oxygen and nitrogen, and not a chemical compound.

13. The air contains water vapor in varying amounts, — carbon dioxide in about three parts in ten thousand, and very small quantities of ammonium compounds and of solid substances.

CHAPTER XIII.

THE COMPOUND OF HYDROGEN AND NITROGEN. (AMMONIA.)

THE not-metallic elements which have been discussed (oxygen and chlorine) unite with hydrogen to form water and hydrogen chloride respectively. Naturally, then, in studying nitrogen we ask if it also will combine with hydrogen, and if it does, what the nature of the resulting compound is.

Direct Union of Nitrogen and Hydrogen Difficult. The chemically indifferent nature of nitrogen becomes apparent when we make the attempt to bring about its union with hydrogen. If an electric spark is passed through a mixture of hydrogen and nitrogen no explosion takes place, and even if the operation is continued for some time the elements will be found to have united only in the smallest quantity. The direct union of the elements, therefore, affords us no practical method for the preparation of the hydrogen compound of nitrogen, but we can easily obtain it by other means.

The Preparation of Pure Ammonia. Concentrated commercial ammonia water is placed in a flask, so arranged that any gases evolved can be passed through a tube containing pieces of quicklime, to a vessel containing mercury, in which is placed a glass tube, closed at one end, filled with the same liquid, and inverted.* The quick-lime is introduced for the purpose of drying the gas which passes off as soon as the flask is gently warmed. This gas can be collected above the mercury in the inverted tube.† [49]

Properties of Ammonia. The gas isolated in this way is termed ammonia. It has a specific gravity, air = 1, of .589; and, being relatively lighter than air, it can be collected by passing it through a long tube extending up to the bottom of an empty jar held mouth downward. This method is convenient if absolutely pure ammonia is not required. Ammonia becomes liquid at a temperature of − 40° under a pressure of one atmosphere. One cubic centimetre of water is capable of dissolving 813 cubic centimetres of ammonia, and it is owing to its extreme solubility in water that we are compelled to collect the gas over mercury.

Commercial ammonia is simply a solution of ammonia gas in water; we have learned (page 16) that gases are less soluble the higher the temperature of the liquid in which they are dissolved. Ammonia gas is, therefore, evolved when ammonia water is heated, and the above experiment for the isolation of ammonia is based upon this fact.[50]

Decomposition of Ammonia. The first question which we must decide in regard to ammonia is whether the gas is an element or a compound. A method of decomposition suggests itself similar to those employed in the study of water and of hydrogen chloride; i.e., to see whether ammonia is decomposed by sodium or potassium. These experiments are not so easily carried out in this case as they were with the substances previously studied, for if sodium or potassium is placed in pure

* Arrangement as in the collecting of gases in the eudiometer tube (see page 24, and Experiment 10 of Laboratory Appendix).

† If the generated gas is allowed to escape freely for a little while before an attempt is made to collect it, the air will all have been expelled from the apparatus so that the pure product will be obtained. If this precaution is not taken, the inverted tube will, of course, be partly filled with air.

ammonia gas at ordinary temperatures no appreciable change takes place. A different result is obtained when the metals are heated; but as hot potassium and sodium are dangerous to handle, we must look for some other metal which, while decomposing ammonia as readily as the two just mentioned, will be easily handled and safe for experimentation. Such a metal is found in magnesium.

Magnesium is a metal with a bright white lustre. When exposed to the air it gradually oxidizes. Unlike sodium or potassium, it does not attack cold water, liberating hydrogen. Magnesium is generally brought into the market in the form of a narrow ribbon, or as wire; the powdered metal is also used to a considerable extent. When ignited in the air, magnesium burns with a brilliant white flame. It should be prevented from tarnishing by being kept in a well-stopped bottle.

Decomposition of Ammonia by Means of Magnesium. A tube of so-called infusible gas, partly filled with pieces of magnesium ribbon, is connected at one end with an apparatus which will generate dry ammonia (Experiment 49), and at the other end with a vessel of water in which is inverted a small bottle filled with water. All air is first expelled from the apparatus by means of a slow current of ammonia; and then, without interrupting the current, the magnesium is gently heated by means of an ordinary burner. The metal soon becomes coated with a grayish powdery crust, while the gas which passes on is collected in the inverted flask over water. When all the liquid has been expelled from the inverted bottle, the latter should be removed, and a lighted taper applied to its mouth. The gas which has been collected burns with a colorless flame, and has the properties which we have learned to associate with hydrogen.* [51]

* Ammonia extinguishes a lighted taper, but does not take fire; hydrogen extinguishes a lighted taper which is plunged into it, but itself takes fire at the point of contact with oxygen. Another distinction between the gas which is here collected and the ammonia from which it is generated is, that ammonia is very soluble in water, while this gas is not.

The Hydrogen which is liberated comes from the Ammonia. We have already seen (page 23) that hydrogen is not formed from the metals which act upon water or hydrochloric acid, but that it owes its origin to a decomposition of those compounds, by means of which the metals take the place of (substitute) the hydrogen. It seems, therefore, scarcely necessary to advance similar proofs that the hydrogen generated by the action of magnesium on ammonia does not come from the magnesium, but does come from the ammonia. Ammonia, therefore, contains hydrogen, which is liberated from it by the action of magnesium, just as the same gas is generated by the action of metals on hydrogen chloride or water. It now remains to be ascertained with what element the hydrogen in ammonia is united.

Isolation of the Nitrogen in Ammonia. For this purpose a bottle is filled with a concentrated solution of common salt, and inverted over a vessel containing the same fluid. This bottle is then completely filled with chlorine generated as described in Experiment 21 of the Appendix. It is then tightly closed with the thumb, and transferred to a vessel containing concentrated ammonia solution, in which it is placed mouth downward. A change immediately sets in, dense white fumes fill the flask, the color of the chlorine disappears, and at the same time the liquid rises in the bottle. When the reaction is entirely at an end, the bottle is again closed with the thumb, and transferred to a vessel containing diluted hydrochloric acid, in which it is opened and allowed to stand for some time. It can now be removed with its mouth closed, then placed upright and opened. The gas which remains is colorless and odorless. A lighted taper placed in it is extinguished; in short, it is nitrogen.* [52]

* The chlorine completely decomposes a portion of the excessive ammonia, forming hydrogen chloride with the hydrogen, and leaving nitrogen. The action of this not-metal on ammonia is, therefore, the reverse of the action of the metal (magnesium).

Parallelism between Ammonia, Hydrogen Chloride, and Water. The foregoing experiments have clearly demonstrated that a metal (magnesium) removes the nitrogen from ammonia and liberates hydrogen, while a not-metal (chlorine) removes *hydrogen* and liberates *nitrogen.* Ammonia, therefore, is a compound of *nitrogen* and *hydrogen;* and its behavior, so far as we have studied it, has been exactly parallel to hydrogen chloride and water.

Relative Volumes of Hydrogen and Nitrogen in Ammonia. The next step in our investigation must be to ascertain the relative *volumes* of hydrogen and of nitrogen in ammonia. This is not so simple an operation as it was with hydrochloric acid or with water.

In order to discover the proportion of nitrogen in a given volume of ammonia, we resort to a modification of the experiment which was based upon its decomposition by chlorine (see page 91, and Experiment 52 of the Laboratory Appendix). In place of the bottle we can substitute a long glass tube, closed at one end, and divided into three equal parts by means of rubber rings. This tube is carefully filled with pure chlorine by the same means that were used in the previous experiment. Then, by successive immersions of the open end in concentrated ammonia solution and in diluted hydrochloric acid, the isolation of nitrogen can be effected as before.* The tube should now be transferred to a deep cylinder of pure water, and lowered therein to the point where the liquid without and within occupies the same level; † it

* As we shall see in the next chapter, hydrochloric acid is capable of uniting with ammonia to form a new compound (ammonium chloride). This acid is, therefore, removed by the surplus of ammonia solution from the nitrogen which is left. Afterward the flask is placed in diluted hydrochloric acid to absorb any excess of ammonia remaining with the gas which has been separated. Diluted sulphuric acid would accomplish the same result.

† The tube must be plunged far enough into the water so that the level of the liquid without and within is the same, because by this

will then be seen that the nitrogen which has been isolated occupies exactly one-third of the total volume of chlorine with which we started.[58] The logical interpretation of this result is as follows: —

We have seen that chlorine unites with an equal volume of hydrogen to form hydrogen chloride (see page 42). In the above experiment, the hydrogen which combined with chlorine came from the ammonia; and as the volume of the chlorine was equal to the capacity of the tube, the volume of hydrogen which was contained in the decomposed ammonia must also have been equal to the capacity of the tube. The nitrogen must have come from that portion of the ammonia which gave up its hydrogen; and as the volume of nitrogen remaining is equal to one-third the total volume of chlorine, it follows that this quantity of nitrogen must have been combined with a volume of hydrogen equal to that of the chlorine which filled the tube. This tube was divided into *three* equal parts in the beginning, and of these three parts *one* is left as nitrogen, so that —

One volume of nitrogen combines with three volumes of hydrogen to form ammonia.

Total Volume of Hydrogen and Nitrogen produced by the Decomposition of Ammonia. A final proof of the volumetric composition of ammonia is easily obtained by the following experiment: —

About 25 cubic centimetres of ammonia gas are enclosed in the eudiometer tube (see page 24), the volume being accurately measured and recalculated to the standard conditions. Electric sparks are now passed through from a battery with an induction coil. These sparks effect the decomposition of the ammonia into nitrogen and hydrogen; and as this goes on the volume of the gas will increase until it reaches 50 cubic centimetres (recalculated), at which point it becomes stationary. If the mouth of the eudiometer is now closed with the thumb, the tube inverted, and a

means we place the remaining nitrogen under atmospheric pressure, as was the chlorine with which we started.

lighted taper applied, the gas will burn. This shows that hydrogen has been liberated, the nitrogen being left uncombined.* [54]

We have already seen that in ammonia three volumes of hydrogen are united with one volume of nitrogen, but we have not seen what volume of ammonia is produced by the union of three volumes of hydrogen with one of nitrogen. The experiment which we are now considering furnishes this evidence, for the 50 cubic centimetres of hydrogen and nitrogen which we have obtained by the decomposition of 25 cubic centimetres of ammonia in the eudiometer tube must contain —

3 × 12.5 cubic centimetres of hydrogen	= 37.5	cubic centimetres.
12.5 cubic centimetres of nitrogen	= 12.5	cubic centimetres.
The total	= 50	cubic centimetres.

These 50 cubic centimetres of the mixed gases are, however, obtained by the breaking down of —

2 × 12.5 cubic centimetres of ammonia = 25 cubic centimetres.

If we reduce all the above figures to simple numbers by dividing by 12.5 we have as a result —

Two cubic centimetres of ammonia produce three cubic centimetres of hydrogen and one cubic centimetre of nitrogen; or, in general terms, two volumes of ammonia are decomposed to form three volumes of hydrogen and one volume of nitrogen. Of course the reverse, that two volumes of ammonia are formed from three volumes of hydrogen and one of nitrogen, must also be true.

Comparison of the Volumetric Composition of Hydrogen Chloride, Water, and Ammonia. If what we have just learned be compared with the results obtained with hydrogen

* Nitrogen does not combine with mercury under the above circumstances, as might be supposed from the action of ammonia on magnesium. This can easily be shown by enclosing some nitrogen in the eudiometer tube and passing electric sparks through the gas, according to the method shown above. The volume of gas will not be altered by this means.

chloride and water, we shall discover the following remarkable regularity: —

One volume of hydrogen combines with one volume of chlorine to form two volumes of hydrogen chloride.

Two volumes of hydrogen combine with one volume of oxygen to form two volumes of water vapor.

Three volumes of hydrogen combine with one volume of nitrogen to form two volumes of ammonia.

Definite Composition of Ammonia. As a result of our investigations with ammonia, we have seen that this substance, like all of the chemical compounds which we have encountered, has a *definite composition.* (See page 55.) The specific gravity of nitrogen is 14 if hydrogen = 1; or, in other words, if a volume of nitrogen weighs 14 grams, the same volume of hydrogen weighs one gram. Three volumes of hydrogen (the amount necessary to form ammonia with one volume of nitrogen) would, therefore, weigh three grams, so that—

Fourteen parts by weight of nitrogen combine with three parts by weight of hydrogen to form seventeen parts by weight of amomnia.

Changes of Energy attending the Decomposition and Formation of Ammonia. In order to decompose ammonia into hydrogen and nitrogen, it was necessary for us to add energy. But it is evident, from the ease with which this decomposition was accomplished by means of the electric spark, that the amount of energy required for a quantity of ammonia containing a given weight of hydrogen is not as great as in the case of an equivalent amount of hydrogen chloride or water. In chemical language, ammonia is *less stable* than the latter two substances. Although it is not practically possible to cause nitrogen and hydrogen to unite directly, these elements must possess chemical energy when in contact; for the compound which they form cannot afterward be decomposed without the addition of energy.

Decomposition of Ammonia and of Water by Chlorine. Reviewing the experience gained in the study of the three hydrogen compounds investigated,* we find that chlorine will easily decompose ammonia, forming hydrogen chloride and nitrogen, and that the same element decomposes water quite slowly, and in the sunlight producing hydrochloric acid and oxygen. This shows us that chlorine and ammonia, and chlorine and water, when in contact, possess chemical energy, which is converted into kinetic energy by the change into hydrochloric acid and nitrogen, and into hydrochloric acid and oxygen. The tendency in these reactions, as in the previous ones which we have studied, is toward a state of more stable equilibrium; the chlorine and water, or the chlorine and ammonia, having more chemical energy than the resulting substances. In these cases, therefore, as in the others, energy is degraded. (See pages 76 and 77.)

Decomposition of Ammonia by Oxygen. Ammonia is also decomposed by oxygen under the proper conditions. A mixture of ammonia gas and oxygen will combine with a weak explosion when ignited. Water and nitrogen are formed by this reaction. It follows

* Hydrogen chloride, water, and ammonia.

that ammonia and oxygen possess more chemical energy than water and nitrogen.[50]

Summary.

1. It is not practically possible to cause the free elements, nitrogen and hydrogen, to unite.
2. Ammonia is decomposed by hot potassium, sodium, or magnesium, while hydrogen is liberated. The hydrogen which is liberated comes from the ammonia.
3. Chlorine easily decomposes ammonia, liberating nitrogen and forming hydrochloric acid.
4. Ammonia is a compound of hydrogen and nitrogen.
5. Three volumes of hydrogen unite with one volume of nitrogen to produce two volumes of ammonia.
6. In comparing chlorine, oxygen, and nitrogen, we find that the volumes of hydrogen which unite with one volume of these elements to form their respective hydrogen compounds are as 1 : 2 : 3.
7. Ammonia has a definite composition by weight. Three parts by weight of hydrogen unite with fourteen parts by weight of nitrogen to produce seventeen parts by weight of ammonia.
8. It takes energy to decompose ammonia, hence energy must be given off in its formation.
9. Chlorine decomposes water to form hydrochloric acid and oxygen, and it also decomposes ammonia to form hydrogen chloride and nitrogen.

CHAPTER XIV.

THE COMPOUNDS OF AMMONIA WITH ACIDS.

The Union of Ammonia with Hydrogen Chloride. If ammonia gas is brought in contact with hydrogen chloride, a change immediately sets in, heat is given off, and dense white clouds of a solid substance are produced, which soon collects on the sides of the vessel in which the reaction has taken place. If exactly equal volumes of ammonia and hydrochloric acid are taken, neither of these substances will remain after the two gases are brought in contact. This fact can easily be proved by the following experiment:—

Two glass tubes of equal size, with one end of each narrowed down, are connected at the narrow ends by means of a short piece of rubber tubing closed at the centre by a pinch-cock. The two glass tubes are filled with mercury and placed mouth downward, side by side, in a mercury trough. Fill one of them with pure ammonia, the other with pure hydrogen chloride. When this has been done, remove the pinch-cock so that a free communication is established between the two. As the union of the two gases takes place, the mercury will rise in the tubes until, provided no air has been admitted, it fills both vessels completely, with the exception of the small space occupied by the solid produced during the reaction. The ammonia and hydrochloric acid have, therefore, without a remainder of either, united to form a solid substance. The bulk of this solid, as compared to that of the original gases, is so small that it can be neglected.[55]

Ammonium Chloride. The solid substance produced by the union of ammonia and hydrogen chloride has the appearance of salt, and in properties and structure can be compared with potassium chloride. We have learned that in ammonia there are three volumes of hydrogen united with one volume of nitrogen, and that in hydrochloric acid there is one volume of hydrogen to every volume of chlorine. If, then, we cause equal volumes of ammonia and hydrogen chloride to unite, we shall have a solid substance which for every volume of nitrogen contains *four* volumes of hydrogen and *one* of chlorine. This solid is termed ammonium chloride.

Resemblance between Potassium Chloride and Ammonium Chloride. We can separate potassium chloride into two substances, — potassium and chlorine. If we remove the chlorine from ammonium chloride, we have as a remainder a compound of one volume of nitrogen with four of hydrogen.* This grouping of the two elements, nitrogen and hydrogen, therefore, in the formation of ammonium chloride, plays the same *rôle* as potassium does in forming potassium chloride. For potassium chloride and ammonium chloride closely resemble each other; and, as the chlorine is the same in both, the remainder, potassium and ammonium, must have a similar effect in determining the character of chemical compounds in which they occur. The difference between potassium and ammonium is that the former has not been found to be decomposable into two or

* This compound is probably capable of only a very brief separate existence under peculiar circumstances. It very soon breaks down into ammonia and hydrogen. It is stable when united with chlorine in ammonium chloride.

more elements, while the latter has. In chemical language, ammonium is termed a *radicle* which can play the part of an element. Such radicles are quite frequent in chemistry, and they display the character of very different chemical elements.

If we represent the one volume of nitrogen in a given amount of ammonia by the letter N, and the three volumes of hydrogen by H_3, we can construct for ammonia a formula, NH_3, which expresses this relationship. If we represent the volume of hydrochloric acid which will unite with the above quantity of ammonia to form ammonium chloride by H Cl, then we can represent the change which takes place when ammonia and hydrogen chloride are brought in contact as follows: —

$$NH_3 + H\,Cl = NH_4\,Cl.$$

Now, if we represent the quantity of potassium which will unite with the amount of chlorine in the above volume of H Cl by the letter K (from the word *kalium*, meaning potassium), then the formula for this weight of potassium chloride will be K Cl, and the relationship between potassium and ammonium chloride is made clear as follows: —

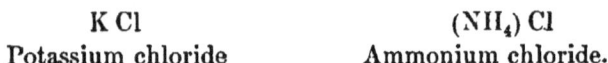

K Cl	(NH_4) Cl
Potassium chloride	Ammonium chloride.

Liberation of Ammonia from Ammonium Chloride. If potassium hydroxide (see page 31) is intimately mixed with ammonium chloride, a change takes place; potassium chloride is formed, and ammonia passes off. During this reaction, water also is produced, as will be evident from the following: —

Potassium hydroxide (caustic potash), we have seen, may be regarded as water in which one-half of the hydrogen has been replaced by the metal potassium. (Page 31.) Ammonium chloride contains, for every volume of nitrogen, four volumes of hydrogen and one of chlorine. When ammonium chloride and potassium hydroxide are brought together, ammonia (which contains one volume of nitrogen to three of hydrogen) passes off. One volume of the hydrogen in ammonium chloride has, therefore, been left behind. In forming potassium chloride, however, the potassium parted with oxygen and one-half as much hydrogen as is necessary to form water. The other half, therefore, was supplied by the ammonium chloride. The change will become more apparent if we once more use formulæ which represent the combining volumes, for then —

$$NH_4\,Cl + KOH = NH_3 + HOH + K\,Cl.*$$

Changes of Energy taking place during the Decomposition of Ammonium Chloride. During the above change, heat is evolved, so that the substances at the right of the sign of equality possess less chemical energy than do those at the left. Chemical energy has been transformed into kinetic energy in passing from the ammonium chloride and potassium hydroxide to ammonia, water, and potassium chloride.† It is partly for this reason that the change takes place spontaneously; yet another factor also comes into play, namely, the volatility of ammonia. It is a matter of repeated experience that decompositions like the above are apt to be complete if a volatile substance can be formed. The reason for this fact cannot be entered into here, for its comprehension requires a somewhat extended knowl-

* In this equation N represents one volume of nitrogen, H_4 four volumes of hydrogen, Cl one volume of chlorine, O one volume of oxygen, and K the amount of potassium which will replace one volume of hydrogen in a quantity of water containing one volume of oxygen. This may be clearer if we substitute the term cubic centimetre for volumes, thus —

N represents one cubic centimetre of nitrogen, etc.

† Ammonium chloride is decomposed not only by potassium hydroxide; but a number of other hydroxides, such as sodium hydroxide, calcium hydroxide, etc., will accomplish the same end. Of course with sodium hydroxide, sodium chloride, and with calcium hydroxide, calcium chloride would be formed.

edge of chemical phenomena. It is, however, not out of place to call attention to another case where the volatility of a compound formed during a chemical reaction determines the completeness of that reaction. We saw that sodium chloride and sulphuric acid produced sodium sulphate and hydrogen chloride. In this case hydrogen chloride was the volatile substance. If, in either of the reactions just referred to, enough water is added to dissolve completely all the ammonia or all the hydrogen chloride which passes off, then, under ordinary circumstances, complete decompositions will not result. Neither ammonia nor hydrogen chloride would then be more volatile than the water in which they are dissolved.

The decomposition of ammonium chloride by bases is a favorite method of preparing ammonia gas for laboratory use. As a general rule, moist calcium hydroxide (slaked lime) is employed for this purpose.[56]

Combination of Ammonia with Acids in General. Ammonia combines with other acids as well as with hydrogen chloride, and in each case the ammonium salt of the corresponding acid is produced. The examples which are of importance to us in this work are as follows:—

Ammonia in contact with nitric acid produces ammonium nitrate.

Ammonia in contact with nitrous acid produces ammonium nitrite.

Ammonia in contact with carbonic acid produces ammonium carbonate.

These three ammonium compounds occur in small amounts in the atmosphere and also in the soil.

If a given quantity of sulphuric acid is brought into contact with one-half as much ammonia as can possibly be taken up by this acid, the primary (acid) sulphate of ammonium is formed. This *primary* sulphate is able to take up a second quantity of ammonia, equal to the first, so as to produce the *secondary* (neutral) ammonium sulphate. There are, therefore, two sulphates of ammonium, just as there are two sulphates of potassium. (See page 59.) Indeed, all the ammonium compounds greatly resemble those of potassium. When brought in contact with soluble bases, like potassium, sodium, or calcium hydroxide, all ammonium salts liberate ammonia exactly as is the case with ammonium chloride.[67]

Changes of Energy in forming Ammonium Compounds. The combination of ammonia with acids to form ammonium compounds is attended with the evolution of heat. The ammonium salts produced, therefore, possess less chemical energy than do the separate ammonia and acids.

Chemical Character of a Solution of Ammonia in Water. The solution of ammonia in water acts very much like a solution of potassium hydroxide. It can neutralize acids to form ammonium salts, just as the latter neutralizes acids to form potassium salts. (See pages 48, 58.) For this reason it has been supposed that the solution of ammonia in water is really ammonium hydroxide; but as all direct proof on this subject is in the contrary direction, we must conclude that ammonium hydroxide does not exist. Further, a belief in ammonium hydroxide is not necessary in order to understand why ammonia solution neutralizes acids, since it is obvious that ammonia gas, when dissolved in water, should be just as capable of union with acids to form ammonium salts as it is in the pure state.

Difference between Ammonium and Potassium Salts. One characteristic sharply distinguishes the ammonium compounds from those of potassium; they are all decomposed by heat, while the ammonia passes off as such if the acid with which it was combined is not volatile. If the acid with which the ammonia is combined is volatile, the ammonia is vaporized with the acid, so that the vapor contains ammonium salts. On the other hand, potassium does not vaporize except at high temperatures; so that, under usual conditions, potassium salts are not volatile.

Summary.

1. A volume of ammonia is capable of uniting with an equal volume of hydrogen chloride to form a salt-like body (ammonium chloride).

2. In ammonium chloride there is combined one volume of nitrogen with four of hydrogen and one of chlorine.

3. The group of substances composed of one volume of nitrogen and four of hydrogen is termed *ammonium*, and its compounds resemble those of *potassium.*

4. Ammonium chloride is decomposed by potassium hydroxide to form potassium chloride, ammonia, and water. Other pronounced bases, such as sodium hydroxide or calcium hydroxide (slaked lime), can also liberate ammonia from ammonium chloride.

5. Heat is evolved during these changes. Therefore the chlorides, the ammonia, and the water which are produced possess less chemical energy than the ammonium chloride and the base, which acted on each other.

6. Ammonia combines with other acids as well as with hydrogen chloride. With the acids it forms ammonium salts, which resemble those of potassium.

There is a primary and secondary sulphate of ammonium.

7. The solution of ammonia in water acts like a base (potassium hydroxide), since it is capable of neutralizing acids to form ammonium salts.

8. By heat, ammonium compounds are decomposed into ammonia and the corresponding acid. If the ammonium salt was that of a volatile acid, the vapors of ammonia and of the acid combine as soon as they become cooler, once more forming ammonium salts.

CHAPTER XV.

THE THEORY WHICH SEEKS TO EXPLAIN THE LAWS OF DEFINITE AND MULTIPLE PROPORTIONS.

The Definite Composition of Chemical Compounds. All the chemical compounds which we have encountered are formed, as we have seen, by the union of definitely related masses of matter. While the number of these compounds has not been great, it has been enough to show us that this definite relationship is not the result of mere chance. The idea of chance is, indeed, entirely eliminated if we extend our view so as to draw upon the experience of others as well as our own. We shall find that all the numberless true chemical compounds with which we are surrounded have this in common, — that they are formed by the interaction of definitely related quantities of the elements. So general has this observation been, that a definite and unalterable composition is regarded as an essential characteristic of a chemical compound.

Meaning of the Term "Definite Composition." What is understood by definite composition has already been explained in the preceding chapters; yet a repetition of a few examples taken from those already studied will develop this meaning with greater clearness.

One volume of hydrogen unites with an equal volume of chlorine to produce hydrogen chloride. If more of either gas than is necessary for this relationship is present before union, then this excess will remain uncombined afterward. Hydrogen chloride is, therefore, formed by the interaction of definitely related volumes of hydrogen and chlorine. A given volume of hydrogen must always have the same weight if it is under the same conditions, so must a given volume of chlorine. The ratio between the weights of equal volumes of hydrogen and chlorine is as 1 : 35.5. Hence one part by weight of hydrogen unites with 35.5 parts by weight of chlorine to form 36.5 parts by weight of hydrogen chloride. The composition by weight of hydrogen chloride is, therefore, a definite one, and so is the relationship between the weights of its constituent parts, hydrogen and chlorine.

Similar considerations have shown us that the ratio between the volumes of hydrogen and oxygen, and hydrogen and nitrogen, which are united in water and ammonia respectively, is a definite one (2 : 1 in water and 3 : 1 in ammonia). The ratio between the weights of *equal* volumes of hydrogen and oxygen is as 1 : 16, and between those of equal volumes of hydrogen and nitrogen as 1 : 14. Therefore the relationship between the weights of hydrogen and oxygen in water is a definite one; i.e., 2 : 16 (2 volumes of hydrogen, 1 volume of oxygen), or in simpler numbers, 1 : 8. Similarly, the relationship between the weights of hydrogen and nitrogen in ammonia is as 3 : 14 (3 volumes hydrogen to 1 volume nitrogen) or 1 : 4.67.

Definite Relationship between the Weights of Chemically Interacting Compounds. That a definite relationship also exists between the weights of chemically interacting bodies is shown by the following examples which we have encountered: —

Sodium attacks water to form sodium hydroxide. Suppose we take 18 grams of water; i.e., the amount of water which is produced by the union of two grams of hydrogen with 16 of oxygen. This quantity of water will be completely converted into sodium hydroxide when exactly one-half the hydrogen (one gram) has been expelled by the sodium. If less sodium is used, then some water will remain unaffected; if more sodium, then some sodium will remain unchanged. Sodium hydroxide, then, has a definite composition; it contains for every 23 parts of sodium, 16 of oxygen and one of hydrogen. It can be shown in the same way that 39 grams of potassium always replace one gram of hydrogen in water, so that potassium hydroxide contains for every 39 parts by weight of potassium, 16 of oxygen and one of hydrogen.

When metals are acted on by dilute acids, a similar definite relationship exists between the weight of metal dissolved, the amount of hydrogen evolved, and the composition of the salt produced. Thus hydrogen chloride, as we have seen, contains 35.5 parts of chlorine to one part of hydrogen. If 36.5 grams of hydrogen chloride are acted upon by enough sodium to replace all the hydrogen, then 23 grams of sodium will be necessary for this decomposition. The sodium chloride which is produced, therefore, contains 23 parts of sodium to every 35.5 parts of chlorine. If potassium is used to act upon the above quantity of hydrogen chloride, 39 parts of potassium replace one part of hydrogen to produce potassium chloride, which salt has 39 parts of potassium to 35.5 parts of chlorine. Parallel results can be obtained with other metals which are attacked by dilute acids. The following table embodies the results developed by a study of the action of a few of the metals on diluted acids and on water: —

One part of hydrogen is replaced by 23 parts of sodium in its action on water.

One part of hydrogen is replaced by 39 parts of potassium in its action on water.

The resulting compounds have the following composition: —

Sodium hydroxide contains 23 parts of sodium to 16 of oxygen and 1 of hydrogen.

Potassium hydroxide contains 39 parts of potassium to 16 of oxygen and 1 of hydrogen.

One part of hydrogen is replaced by 23 parts of sodium in its action on hydrogen chloride.

One part of hydrogen is replaced by 39 parts of potassium in its action on hydrogen chloride.

One part of hydrogen is replaced by 32.6 parts of zinc in its action on hydrogen chloride.

One part of hydrogen is replaced by 12.1 parts of magnesium in its action on hydrogen chloride.

One part of hydrogen is replaced by 28 parts of iron in its action on hydrogen chloride.

The resulting compounds have the following composition: —

Sodium chloride contains 23 parts of sodium to 35.5 parts of chlorine.

Potassium chloride contains 39 parts of potassium to 35.5 parts of chlorine.

Zinc chloride contains 32.6 parts of zinc to 35.5 parts of chlorine.

Magnesium chloride contains 12.1 parts of magnesium to 35.5 parts of chlorine.

Iron chloride contains 28 parts of iron to 35.5 parts of chlorine.

One part of hydrogen is replaced by 23 parts of sodium in its action on sulphuric acid.

One part of hydrogen is replaced by 39 parts of potassium in its action on sulphuric acid.

One part of hydrogen is replaced by 32.6 parts of zinc in its action on sulphuric acid.

One part of hydrogen is replaced by 12.1 parts of magnesium in its action on sulphuric acid.

One part of hydrogen is replaced by 28 parts of iron in its action on sulphuric acid.

The resulting compounds have the following composition: —

Sodium sulphate contains 23 parts of sodium to 16 of sulphur and 32 of oxygen.

Potassium sulphate contains 39 parts of potassium to 16 of sulphur and 32 of oxygen.

Zinc sulphate contains 32.6 parts of zinc to 16 of sulphur and 32 of oxygen.

Magnesium sulphate contains 12.1 parts of magnesium to 16 of sulphur and 32 of oxygen.

Iron sulphate contains 28 parts of iron to 16 of sulphur and 32 of oxygen.

The Equivalent Weights of the Metals. It will be noticed, on examining the above table, that the weight of any given one of the metals which replaces one part of hydrogen in the acids is the same, no matter whether we use hydrochloric or sulphuric acid. The same is true whatever acid we employ. For example, 23 parts of sodium always replace one part of hydrogen in any acid. The relative weights of the metals which are given above are termed their equivalent weights, since they are equivalent in combining power to one part of hydrogen. These equivalent weights can be easily determined by experiment. We have but to weigh a small quantity of the metal whose equivalent weight we wish to determine, dissolve this in an acid, and carefully collect and measure the hydrogen evolved. Knowing the weight of one cubic centimetre of hydrogen,* we can readily discover the volume occupied by one gram, and, knowing the weight of the metal dissolved and the number of cubic centimetres of hydrogen evolved, we can calculate the quantity of metal necessary to give us one gram of hydrogen.[58] The salts mentioned in the above table can also be produced by neutralizing the corresponding acids with the corresponding hydroxides of the metals; but their composition by weight remains unaltered, no matter what has been their method of formation.

An Explanation of the Existence of Fixed Relationship by Weight in Chemical Compounds is found in the Atomic Theory. As the fixed relationship existing between the masses of the combining elements in chemical compounds is not the result of mere chance, it follows

* One cubic centimetre of hydrogen weighs .00009001 gram.

that some reason may be found for this regularity. Why should chemical compounds always show an unvarying composition by weight? Why should not water, for example, at one time contain, for every one part of hydrogen, seven of oxygen, and at another time nine? The scientific world has sought for an explanation of these conditions, and has produced a theory which satisfactorily accounts for all the phenomena encountered. This theory is known as the atomic theory.

The Atomic Theory. Let us suppose that all of the elements are composed of extremely small particles which we will term atoms. Let us further suppose that the weight of an atom of a given element is equal to that of each other atom of the same element, but differs from that of an atom of any other element. The atoms of different elements unite to form the smallest individual group of the compound. Atoms of the same kind unite to produce the smallest individual group of an element. These groups we term molecules, and the agglomeration of molecules forms tangible matter. To make the results of a theory such as this more apparent, we can use an illustration in which certain visible portions of matter will serve as atoms. Let us take two kinds of shot, one kind lead, the other copper, the lead shot each to weigh one gram, the copper shot five grams. If, now, one of the lead is fastened to one of the copper, the resulting combination will contain one part of lead to five of copper. Next, let us continue this process of uniting one lead shot with one copper shot until we have a large number together, say six hundred grams. It is obvious that, if we separate all the copper from all the lead, we shall have five hundred grams of the former to one hundred grams of the latter. In other words, this large mass of combined lead and copper will have the same relationship by weight between the constituent elements as is found in the individual shot. Furthermore, if the above conditions have been carefully adhered to, it will not be possible to have any other relationship between the copper and lead than that of five to one, no matter how many shot we take. Obviously it is not necessary to know how many shot we have in a given mass in order to ascertain the relative weights of the particles of lead and of copper. We need but to separate and weigh all the lead and all the copper to know that the proportion of each is as one to five. While this last separation will tell us what the proportional parts by weight of the constituent metals in each of the combined particles are, it will not tell us how many individual shot are united in each of these unless we know the weight of each. If some one, for example, had substituted lead shot weighing half a gram each for those originally used, and had attached two lead shot to one of copper, the results on separating the lead from the copper would remain the same, five parts of copper to one of lead. In such case we could not ascertain the number of lead shot without counting them.

Difficulty of Determining the Number of Atoms combined in a Molecule. It is in this latter condition that chemistry finds itself. It can take large masses of matter, separate these into their constituent elements, and weigh the latter. It can show that so many parts by weight of one element combine with so many parts by weight of another. It can, therefore, determine that the smallest particles of the compound (the molecules) contain so many parts of one element to so many parts of another; but without some special means of counting, it cannot determine the number of atoms of any of the elements which go to make up those smallest particles. A direct count, as in the case of the shot, is out of the question, for we can neither see nor weigh the individual atoms. For example, we have abundant proof in the preceding work that one part of hydrogen combines with eight parts of oxygen to form nine parts of water. We have the theory that hydrogen and oxygen are formed of atoms, that one or more atoms of hydrogen unite with one or more atoms of oxygen to form the smallest particle (the molecule) of water. We can say that this molecule contains one part of hydrogen to eight of oxygen, but we cannot, with the knowledge which we now possess, determine how many atoms of hydrogen and how many atoms of oxygen there are in this molecule. As a consequence, we cannot, using the knowledge we have so far obtained, ascertain the relative weights of these atoms in this case. We merely know that there is eight times as much oxygen as hydrogen. We may, at the very outset, abandon the hope of determining the absolute weights of the atoms; for, unlike the shot which we used as an illustration, they are too small to be weighed directly. What is true of water is also true of hydrogen chloride, of ammonia, of caustic soda, and of every other compound which we have studied and found to be of definite composition by weight. We can determine the number of parts by weight of each element which enters into these compounds. We can suppose that the cause of this definite relationship is to be found in the fact that these elements are composed of atoms, each having a definite mass. We assume that the combination of these atoms produces molecules of a fixed composition which belongs also to the tangible quantities of the compounds produced by heaping together these molecules, but further than this we cannot go.

Practical Advantages of the Atomic Theory as we have so far developed it. The atomic theory, as we have so far developed it, has, therefore, no particular practical bearing on our understanding of chemistry. It affords an explanation of the law of definite proportions (see page 55); but without it we can just as well go on studying chemical reactions and chemical energy, and determine the relative parts by weight in which the elements combine. If, however, we can find some method of ascertaining the number of atoms combined in the individual molecules of a large number of chemical compounds, we can ascertain the relative weights of the individual atoms. By this means, possibly, we can make clear relationships existing between compounds which would otherwise be concealed. In that case the atomic theory would be of the greatest practical importance.

Thoroughly scientific means of ascertaining the number of atoms united in the individual molecules of many compounds really have been established, and one of the methods, the most important, we will examine in the next chapter. Before we take up this part of the discussion, however, we must examine two other chemical laws which, while not absolutely essential in establishing the atomic theory, are, nevertheless, of great importance in furnishing proof that is based upon correct reasoning.

The Law of Multiple Proportions. The first of these laws is known as the law of multiple proportions. The illustrations of this law which we have studied were found during the discussions of the oxides of sulphur and of sulphuric acid. (See Chapter IX.) We there saw that sulphur is capable of forming two oxides, sulphur dioxide and sulphur trioxide, and the parts of oxygen which unite with one part of sulphur in these two compounds are to each other as 1 : 1.5. If we multiply by two, so as to produce whole numbers, we have —

Two parts of sulphur are united with two parts of oxygen in sulphur dioxide.

Two parts of sulphur are united with three parts of oxygen in sulphur trioxide.

This individual case is but an example of what general experience in chemistry has taught us. Many elements, like sulphur and oxygen, combine with each other in more than one proportion, forming more than one compound. In every instance where one element (A) unites with a second element (B) in more than one proportion, in the series of compounds so produced the proportional parts of B, which combine with a given quantity of A, are to each other in a simple ratio such as 1 : 2, or 2 : 3, or 3 : 4. What is true of compounds composed of two elements is also true of those containing three or more.

Relation of the Law of Multiple Proportions to the Atomic Theory. This law of multiple proportions can readily be explained by the atomic theory, as will be evident if we once more resort to our illustration with the lead and the copper shot. We have a number of lead shot each weighing one gram, and a number of copper shot each weighing five grams. If we fasten one lead shot to one of copper, we shall have produced a mass containing one part of lead to five of copper. Now let us take a second lead shot and attach it to two of copper. It is obvious that the resulting body will have one part of lead to ten of copper, and if we compare the two combinations, we see that the weight of the copper united to one part of lead in the first, is to the weight of copper united to one part of lead in the second, as 1 : 2. The same must be true if we take any number of individuals of the first combination and compare them with any number of the second. The proportion between the weights of the lead and copper in the first will always be as 1 : 5, and in the second as 1 : 10; and the relation between the weight of copper united with one part of lead in the two will always be as 1 : 2.

We can explain the law of multiple proportions in the same way if we use the atomic theory outlined on page 110. We have two oxides of sulphur. In the first we have two parts of sulphur to two of oxygen, and in the second two parts of sulphur to three of oxygen. Let us suppose now that the smallest particles (the molecules) of the first oxide are each formed of one atom of sulphur united to two atoms of oxygen, each molecule containing equal parts by weight of sulphur and oxygen. One atom of oxygen would then weigh half as much as an atom of sulphur. Since any large mass of this sulphur oxide would be formed of these molecules, it would follow that, if we separated all the sulphur from all the oxygen in this large mass, we should obtain equal weights of sulphur and of oxygen. Next, let us return to the molecule and add to it one atom of oxygen. This second molecule would then contain two parts of sulphur to *three* of oxygen, and, obviously, a tangible mass made up of these molecules would also be separable into two parts of sulphur and three of oxygen. The relationship existing between the amounts of oxygen united to a given weight of sulphur in the two oxides of sulphur (sulphur dioxide and sulphur trioxide) is therefore explained by the atomic theory. If this explanation applies to the two compounds which we have discussed, it must also necessarily apply to any other series of compounds which, like the two oxides of sulphur, are formed of two or more elements according to the law of multiple proportions.

The Law of Multiple Proportions cannot help us to determine the Relative Weights of the Atoms. We cannot, however, see and weigh the molecules directly; they are too small. We can only determine the proportional parts in which the elements unite in tangible masses of the compounds. It follows from this that we cannot, by ascertaining these proportional parts alone, absolutely fix the relative weights of the atoms themselves; we can only say that they must be so united in the molecules of the compounds that the weight of all the atoms of one element in one molecule must be to the weight of all the atoms of the second element in that molecule as the weight of one element is to the weight of the other in the large quantity of the compound with which we come in contact. That this is so will be seen by the following:—

We supposed that one atom of sulphur was united to two atoms of oxygen in the molecules of the first oxide of sulphur, and that one atom of sulphur was united to three of oxygen in those of the second. Let us now change this hypothesis to read that one atom of sulphur is joined to four of oxygen in one molecule of sulphur dioxide, and to six in one of sulphur trioxide, one atom of oxygen having just one-fourth of the weight of one atom of sulphur. The latter view of the case would agree with the experimental results exactly as well as the former; for if the molecules of the oxides of sulphur were constructed in this way, sulphur dioxide would have two parts of sulphur united to two of oxygen, and sulphur trioxide two of sulphur joined to three of oxygen. These facts are unalterable, and any hypothesis which we choose to adopt must agree with them.

We must therefore come to the conclusion that, while the relations existing between the proportional parts in which oxygen and sulphur unite are explainable by the atomic hypothesis, the determination of these proportional parts alone cannot possibly help us to ascertain the relative weights of the atoms themselves. The atomic theory, therefore, in the form which we at present have it in this work, has no particular bearing on the systematic study of chemistry. This conclusion we have already reached in the law of definite proportions.

The Equivalent Weights of Elements. One other series of facts in support of the atomic theory is found by comparing a number of the compounds, each of which contains the same element united to others, which latter are however different in each one of the series. This comparison has already been instituted in the table which was constructed at the beginning of this chapter (pages 108 and 109).

On consulting this table we find that —

23 parts of sodium,
39 parts of potassium,
32.6 parts of zinc,
12.1 parts of magnesium,
28 parts of iron,
} are united with

16 parts of oxygen and 1 of hydrogen in the hydroxides of the above metals, and with 35.5 parts of chlorine in the chlorides of the above metals.

Furthermore, as we have seen, 8 parts of oxygen are united with one part of hydrogen in water. This table shows us that 2 times 8 parts of oxygen are combined with 23 parts of sodium in sodium hydroxide, and with 39 parts of potassium in potassium hydroxide. Also on page 109 we saw that 4 times 8 parts of oxygen are found together with 23 of sodium and 39 of potassium in potassium sulphate.

The examples which have been cited have been confined to the few elements encountered during the progress of this work, but we should have arrived at similar results had we used any other ones for our basis of study. If, therefore, we define as the equivalent weight of an element that part of this element which will combine with one part of hydrogen, or which will replace one part of hydrogen in a chemical compound (see page 110), we shall arrive at the following rule: —

All true chemical compounds are produced by the union of the equivalent weights, or of simple multiples or sub-multiples of the equivalent weights, of the elements.

The Formation of Compounds from the Equivalent Weights of the Elements is in accordance with the Atomic Theory. The equivalent weight of an element is, therefore, a constant which accompanies it throughout its combinations with other elements. It remains while apparently all of the other characteristics of the element (i.e., color, malleability, ductility, etc.) have been lost in the formation of a compound differing in all its properties from the elements of which it is composed.

This result is entirely in accordance with the atomic theory; for, if the elements are formed of atoms, these atoms must always retain the same relative weights, no matter how varied are the compounds formed by them. If an atom of sodium weighs

23 times as much as an atom of hydrogen, and if one molecule of water is acted on by sodium in such a way that an atom of sodium replaces one of hydrogen, then 23 parts of sodium will take the place of one part of hydrogen, no matter how large a quantity of water or of sodium is taken. Furthermore, if one atom of potassium weighs 39, one atom of zinc 32.6, one of iron 28, times as much as one of hydrogen, then potassium, zinc, or iron, must retain these relative weights, no matter in what compound they are encountered.

The Relative Weights of the Atoms are Simple Multiples or Sub-multiples of the Equivalent Weights. At first glance it would seem that the equivalent weights might be considered as the relative weights of the atoms themselves, but a little reflection will show us that such a conclusion would be arbitrary and without warrant. In the above paragraph, for example, we have assumed that one atom of sodium replaces one atom of hydrogen in water. If this were true, the relative weights of the atoms of hydrogen and of sodium would be the same as the equivalent weights; namely, 1 and 23. With equal right we might suppose, however, that two atoms of hydrogen are replaced by one of sodium, in which case the relative weights of the atoms of hydrogen and sodium would be 1 and 46, and the same considerations would show us that the relative atomic weights of iron and hydrogen might be 1 and 28, or 1 and 56, etc. The equivalent weights of the various elements may possibly represent the relative weights of the atoms themselves, but they may also be some simple multiple or sub-multiple of those atomic weights. While the fact that all true chemical compounds are produced by the union of the equivalent weights, or of simple multiples or sub-multiples of the equivalent weights of the elements, is an added support to the atomic theory, the determination of these equivalent weights does not, without further experimental aid, help us to decide upon the true relative weights of the atoms themselves.

After a consideration of all the arguments which have been advanced, the question naturally presents itself: Are no experimental means at our disposal which will enable us to determine the relative weights of the atoms? Must the atomic theory always be to us simply an explanation of the reason why certain regularities in the composition of chemical compounds are encountered, without having any practical bearing in ascertaining the number of atoms which are combined in the chemical compounds, or the relative weights of the atoms themselves? An attempt to answer these questions involves a more minute consideration of the laws underlying the combining volumes of gases, and is, therefore, reserved for a separate chapter.

Summary.

1. All chemical compounds are formed by the union of definitely related masses of matter.

2. A definite relationship also exists between the weights of chemically interacting bodies.

3. The quantity of any metal which is capable of liberating one part of hydrogen from acids is termed the equivalent weight of that metal. A given weight of a metal when acted upon by certain dilute acids always liberates the same quantity of hydrogen.

4. The salts formed by the substitution of hydrogen in acids have a definite composition by weight. This composition is constant, no matter what is the method of their formation.

5. The fixed relationship existing between the combining weights of the elements is explainable by means of the atomic theory.

6. The atomic theory supposes that all the elements are made up of extremely small particles termed atoms. The weight of an atom of a given element is equal to that of each other atom of the same element, but differs from that of an atom of any other element.

7. The atoms of different elements unite to form the smallest individual group of a compound. This group is termed a molecule.

8. While the fact that a given chemical compound always has a definite composition is explainable by means of the atomic hypothesis, the discovery of the relative weights of the different elements which form that compound cannot determine for us the relative weights of the atoms themselves.

9. Chemical compounds are produced by the union of the equivalent, or of simple multiples or sub-multiples of the equivalent, weights of the elements.

CHAPTER XVI.

MODERN THEORY OF THE NATURE OF GASES.

The Relation between Specific Gravities of Gases and their Molecular Weights.

A CONSIDERATION of the relative parts by weight in which the atoms unite cannot provide us with a satisfactory means of determining the relative weights of the atoms themselves. But a combination of the combining weights with the combining volumes and with the specific gravities of the gaseous elements and compounds furnishes us with a means of deciding what multiples or sub-multiples of those combining weights we must regard as representing the true relative weights of the atoms. In order to comprehend how this combination of the three constants (combining weights, combining volumes, and specific gravities of gaseous elements and compounds) has led to the desired result, we must first understand the prevailing belief as to the physical nature of a gas. This prevailing belief is in accordance with the atomic theory.

The Kinetic Gas Theory. A gas is at the present time regarded as a form of matter, the individual particles of which (molecules) are so far separated that each is capable of motion independently of the others.

That the particles of a gas must be comparatively far apart is proven by the fact that they occupy a space many hundred times greater than that of the liquids from which they are formed by heating. These particles are at such a distance from each other that they are not subject to those mutual influences (cohesion, etc.) which are present in liquids and solids. They are, therefore, acted upon only by the attraction which all masses exert on each other at a distance (gravitation). As a consequence, the individual molecules attract each other only in a very slight degree. Accordingly, the small particles of which a gas is composed are continually in motion, and, following the ordinary laws of mechanics, move in right lines until they collide with some other molecule of their own kind, or with the walls of the enclosing vessel, when, being perfectly elastic, they rebound. The pressure exerted by a gas is, therefore, due to the continual impacts given by its molecules upon the sides of the retainer. This pressure must consequently increase with the number of molecules and with the mass and velocity of each molecule. The energy possessed by a moving body is measured by its mass multiplied by the square of its velocity and divided by two $\left(\frac{Mv^2}{2}\right)$. Therefore, if we multiply the mass of each individual molecule in a given volume of the gas by one-half the square of its velocity, we shall obtain the energy of motion possessed by that molecule; and the sum of all these products will, obviously, be a quantity proportional to the pressure of the gas. In discussing the atmosphere, we learned that gases expand $\frac{1}{273}$ of their volume for each rise of one degree in temperature (law of Gay Lussac, see page 69). However, if the volume of the gas be kept constant while the temperature is raised, then the increase of the pressure of the gas is found to be proportional to the increase in the temperature (the latter being calculated from $-273°$). In this volume of gas the *mass* remains constant, so that it follows that the *temperature* is proportional to the square of the velocity of the molecules.

Let us suppose that we have two equal spaces, each filled with a different gas, under the same temperature and pressure. From what has been said above, the sum of the kinetic energy possessed by the particles in one of the gases will be equal to the sum in the other. We will assume that *in these equal gas volumes there are equal numbers of molecules* (hypothesis of Avogadro). It will follow that the kinetic energy $\left(\frac{Mv^2}{2}\right)$ belonging to each individual particle in either of the gases averages the same. If both gas volumes are brought into communication, they will mix without suffering either a change in temperature or pressure, provided always that they exert no chemical action on each other. In the mixture so produced, therefore, there is also on the average the same amount of kinetic energy belonging to each individual molecule.

In Equal Volumes of Gases, under like Conditions, there are Equal Numbers of Molecules. Assuming that *in equal volumes of gases, under the same temperature and pressure, there are equal numbers of molecules*, if we take into consideration only the simple laws of mechanics, we can understand why two gases can be mixed without a change in temperature or total volume. We can also find a reason for like changes in volume caused by like changes in temperature (law of Gay Lussac), and for the fact that all gases expand alike and are altered alike in volume by like changes in pressure (law of Boyle, see page 68).

The matter assumes an entirely different aspect, however, if we do not assume that in equal volumes of gases there are equal numbers of particles under like conditions of temperature and pressure. For example, suppose that one of the above gas volumes contains twice as many molecules as the other. Then, since the pressure exerted by both gases is alike, and the sum of the kinetic energy of all the particles in the one must be equal to that sum in the other, each of these particles will possess only one-half the kinetic energy which belongs to a particle of the other gas. It would then be difficult to understand how the equality of temperature and pressure could be preserved after the two gases are mixed. For we should have in this mixture particles possessing twice as much kinetic energy as others, and we could see no reason why, owing to frequent collisions, these should not impart some of their energy to those with which they come in contact. If this were to occur, however, we could scarcely maintain the former equality of temperature and pressure, since both of these are proportional to the kinetic energy possessed by the molecules. Avogadro's hypothesis, that in equal volumes of gases there are equal numbers of molecules, is, therefore, the only one in accordance with the laws of mechanics. With the establishment of this hypothesis we have a method of determining the relative weights of the molecules of gaseous substances.

The weight of a given volume of gas is obviously the sum of the weights of the molecules of which it is composed. If, now, we weigh equal volumes of a series of different gases, all under the same conditions as to the temperature and pressure, the weights so obtained will have the same relationship to each other as do the weights of the individual molecules, because in equal volumes of gases there are equal numbers of the latter.

Relation between the Molecular Weights and Specific Gravities of Gases. Let us suppose that we have equal volumes of two gases, the molecules of which weigh x and y respectively. If the number of molecules in the first volume is n, then the number in the second will also be n, and —

$$nx : ny :: x : y$$

but nx and ny represent the weights of the two gases * (w and w'), so that —

$$w : w' :: x : y.$$

The weights of equal volumes of various gases, when compared with the weight of an equal volume of some other gas taken as a standard, represent the *specific gravities* of those gases. From what has gone before, we can establish the following rule : —

* If all of the molecules of a given gas volume are of the same kind, then the sum of the weights of the individual molecules can be represented as a product, where n represents the number of molecules.

The molecular weights (the relative weights of the individual molecules) *of gases are to each other as the specific gravities of those gases.*

Standard for Specific Gravities of Gases. It is generally customary to compare all specific gravities of gases with that of air as unity; i.e., to weigh gas volumes which are equal to that occupied by one gram of air. For our purpose, however, it will be better to use *hydrogen* as a standard, and to understand by the term specific gravity of a gas, the relative weights of volumes equal to the space occupied by the unit weight of hydrogen.

EXAMPLES: — One gram of hydrogen occupies 11.111 + litres at 0° and 760mm. pressure.

11.111 + litres of oxygen weigh 16 grams,
11.111 + litres of chlorine weigh 35.5 grams,
11.111 + litres of nitrogen weigh 14 grams,
11.111 + litres of hydrogen chloride weigh 18.25 grams,
11.111 + litres of water weigh 9 grams,
11.111 + litres of ammonia weigh 8.5 grams,

and the specific gravities are as follows: —

Hydrogen	= 1.
Oxygen	= 16.
Chlorine	= 35.5.
Nitrogen	= 14.
Hydrogen chloride	= 18.25.
Water	= 9.
Ammonia	= 8.5.

These specific gravities, as we have seen, bear the same relationship to each other as do the molecular weights of the respective gases. *We need then but to decide upon the molecular weight of one of them in order to have given to us the relative molecular weights of all of the others.* Such a decision has been reached, and the reasons which have led to it will be outlined in the next chapter.

Summary.

1. The individual particles (molecules) of a gas are so far separated that each is capable of motion independently of the others.

2. The pressure exerted by a gas is due to the impacts of its molecules upon the sides of the container.

3. The sum of the kinetic energy of all the molecules is proportional to the gas pressure.

4. In equal volumes of gases, under the same conditions of temperature and pressure, there are equal numbers of molecules.

5. The weights of equal volumes of gases are to each other as the molecular weights.

6. The molecular weights are to each other as the specific gravities.

7. The specific gravities of gases are usually ascertained with air as a unity. In our chemical work it is better to select hydrogen as the standard.

8. If we decide upon the molecular weight of one gas, we can, by determining their specific gravities, discover the molecular weights of other gases.

CHAPTER XVII.

THE DETERMINATION OF ATOMIC WEIGHTS BY USE OF THE SPECIFIC GRAVITIES OF GASES.

In the preceding chapter we learned that, because equal volumes of gases contain equal numbers of molecules, the relative weights of equal volumes of gases must bear the same relationship to each other as the molecular weights. The determination of various molecular weights, therefore, becomes a comparatively easy matter, provided we can determine the molecular weight of some one element which we can select as a standard, and with which we can compare the other gaseous substances. In order to do this we must examine more closely into the lessons taught us by the study of the combining volumes of elementary gases.

The Molecules of Hydrogen consist of Two Atoms. *One* volume of hydrogen unites with *one* volume of chlorine to produce *two* volumes of hydrogen chloride. Let us now suppose that we have a volume of hydrogen which contains one hundred molecules. From what we have learned, it follows that an equal volume of chlorine also contains one hundred molecules. These gas volumes we will represent by squares, as follows: —

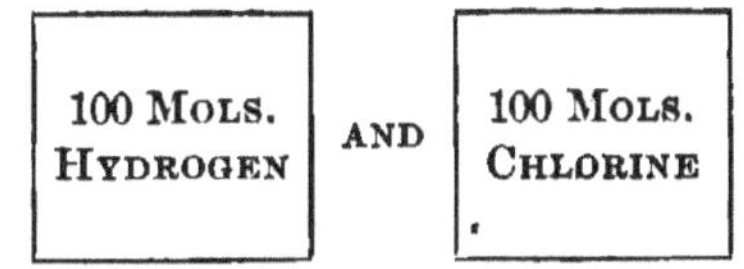

1 volume hydrogen. 1 volume chlorine.

The above volumes of hydrogen and chlorine are capable of uniting to form a quantity of hydrogen chloride equal to twice the volume of hydrogen: —

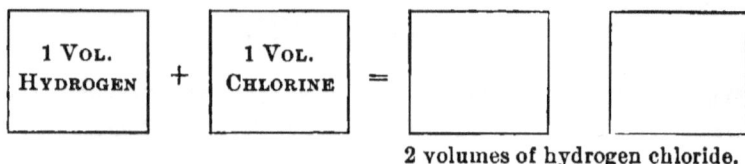

1 Vol. Hydrogen + 1 Vol. Chlorine = [ ] [ ]

2 volumes of hydrogen chloride.

Now, each volume of hydrogen chloride must contain *as many molecules as an equal volume of hydrogen or of chlorine;* i.e., 100 molecules, so that —

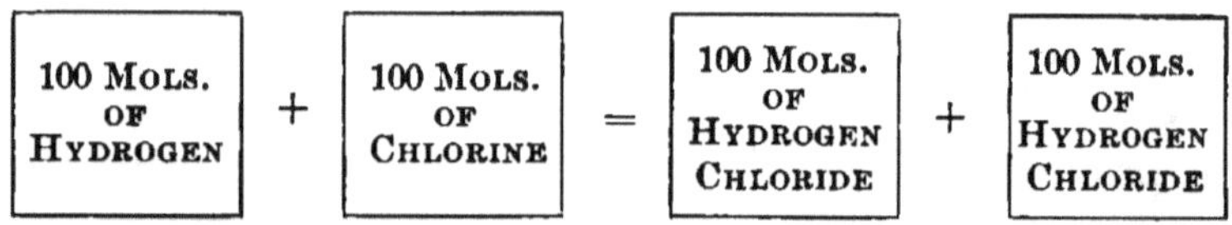

It follows from this, however, that 100 *molecules* of hydrogen must have taken part in the formation of 200 *molecules* of hydrogen chloride, so that each molecule of hydrogen *must have been divided into two parts.* The molecule of hydrogen thus contains *at least two atoms.* The same of necessity must be true also of chlorine, for the 100 molecules of chlorine have also taken part in the formation of 200 molecules of hydrogen chloride. What is true of the above gas volumes is obviously true of any other equal volumes of hydrogen and chlorine, for in equal volumes there are equal numbers of molecules. The above reasoning does not exclude a theory that a molecule of hydrogen *may* contain more than two atoms; but such a view has no great probability, because in the formation of the various hydrogen compounds we encounter no instance in which one volume of hydrogen enters into the production of *more* than two volumes of the compound gas.

EXAMPLES: — Two volumes of hydrogen and one of oxygen produce *two* volumes of water vapor.

Three volumes of hydrogen and one of nitrogen produce *two* volumes of ammonia.

The Molecular Weight of Hydrogen can be taken as the Standard for Measuring other Molecular Weights. It seems reasonably certain, therefore, that one molecule of hydrogen is formed of two atoms. Its molecular weight, then, is twice its atomic weight. We have no accurate means of determining the absolute weights of the atoms, but for our purposes this is not necessary. The relative weights answer all purposes; and we need only select the weight of an atom of some one element as a standard, and measure all others by this.* Hydrogen is, relatively, the lightest of known elements.

It is convenient, therefore, to fix upon the weight of an atom of hydrogen as unity, and to say that other atoms weigh 16, 14, 35.5, etc., times as much as one atom of hydrogen. If the atom of hydrogen weighs 1, then the *molecule* of hydrogen weighs 2.

Determination of the Molecular Weights of Gases. We are now in a position to determine the relative molecular weights of other gaseous elements and compounds; for we need but select a volume of hydrogen which weighs two grams, and then weigh off volumes of the other gases equal to that volume of hydrogen. Their relative weights will represent the relative weights of the molecules.

* The object of measuring anything is simply to determine its relation to other things. We fix upon some arbitrary standard (a foot, a metre, an inch) as a unit, and measure all other lengths by this. The same is true of weights — the unit of weight (a grain, a gram, a kilogram) is one of arbitrary selection.

If n represents the number of molecules in 2 grams of hydrogen, then $\frac{2}{n}$ will be the weight of each individual molecule. If x, y, z represent the weights of volumes of other gases equal to that of 2 grams hydrogen, then $\frac{x}{n}$, $\frac{y}{n}$, $\frac{z}{n}$, will be the relative weights of the individual molecules of each of those gases, since n remains the same; and $2 : x : y : z :: \frac{2}{n} : \frac{x}{n} : \frac{y}{n} : \frac{z}{n}$. Hence, if 2 represents the weight of one molecule of hydrogen, the relative weights of the other molecules are represented by x, y, and z; *i.e.*, by their specific gravities. Obviously the same result will be obtained if we begin, not with two grams of hydrogen, but with two units of any other standard of weight, so that we can establish the following rule:—

If hydrogen is placed at 2, the specific gravities of all other gases (which are measured by these two parts of hydrogen) are the same numbers as their molecular weights. This rule is illustrated by the following table: —

Hydrogen = 2.

GAS.	SPECIFIC GRAVITY.	MOLECULAR WEIGHT.
Chlorine	71.	71.
Oxygen	32.	32.
Nitrogen	28.	28.
Hydrogen chloride	36.5	36.5
Water (gas)	18.	18.
Ammonia	17.	17.

Determination of the Maximum Atomic Weights of the Elements. By means of these results we obtain a method for determining the *maximum* numbers which can represent the relative weights of the atoms of chlorine, oxygen, and nitrogen. This conclusion is evident if we analyze the figures a little more closely.

The weight of a molecule of water is 18 if the molecule of hydrogen is 2. Our previous work has shown us that in 18 parts of water there are 16 of oxygen and two of hydrogen. Hence one molecule of water contains 16 parts of oxygen and two of hydrogen. Obviously, then, one atom of oxygen cannot weigh more than 16 times as much as one atom of hydrogen; for, if its relative weight were supposed to be greater than this, one molecule of water would contain a fraction of one atom of oxygen. We cannot with equal certainty state that an atom of oxygen may not have a relative weight of less than 16, for it is easily conceivable that a molecule of water may contain more than one atom of oxygen. Such a supposition, however, would be entirely without experimental backing. We have never encountered any one of the numerous compounds of oxygen (which we can obtain as gases, and the molecular weights of which we consequently know) in which we have *less* than 16 parts of oxygen in each molecule. If, therefore, we place the weight of one atom of hydrogen at unity (one-half the *molecular weight*), we have every reason to suppose that one atom of oxygen weighs *16 times as much as that unit.* In one molecule of water we therefore have *two atoms of hydrogen and one of oxygen.*

The relative weight of a molecule of hydrogen chloride is 36.5. This molecule contains one part of hydrogen and 35.5 of chlorine. The maximum atomic weight of chlorine is, therefore, 35.5; and as we have never encountered less than this quantity of chlorine united with one part of hydrogen or its equivalent in any gasifiable chlorine compound, we must conclude that 35.5 is the minimum. One atom of chlorine, therefore, weighs 35.5 times as much as an atom of hydrogen, and in one molecule of hydrogen chloride we have *one* atom of hydrogen and *one* atom of chlorine.

The relative weight of a molecule of ammonia is 17; this molecule contains three parts of hydrogen to 14 of nitrogen. The maximum atomic weight of nitrogen is, therefore, 14, and for reasons similar to those given with oxygen and chlorine, its minimum is also 14. One molecule of ammonia, then, contains *three atoms* of hydrogen and *one atom* of nitrogen.

The above table now shows us that the relative molecular weights of the elements, chlorine, oxygen, and nitrogen, represent numbers which are twice as large as the relative atomic weights of those elements as determined from their hydrogen compounds. Therefore, the molecules of these elementary gases contain at least *two atoms.* We can now construct the following table:* —

The weight of an atom of hydrogen is placed = 1.

Elements.	Atomic Weight.	Molecular Weight.
Hydrogen . . .	1.	2.
Chlorine	35.5	71.
Oxygen	16.	32.
Nitrogen	14.	28.

Chemical notation can be very much simplified if we represent the atoms of each of the elements, with the atomic weights of which we are acquainted, by means of certain symbols. The system universally adopted is one which uses the first letter of the English, Latin, or Latinized name of the element in question; i.e., H represents *one atom* of hydrogen, O *one atom* of oxygen, etc.† The structure of the molecules of the elements is then expressed by writing after the symbol a subscript number, representing the number of atoms in the molecule; e.g., H_2 represents one molecule of hydrogen, O_2 one molecule of oxygen, N_2 one molecule of nitrogen. The molecules of compounds are represented by writing the symbols of the elements entering into their formation side by side, with the subscript numbers rep-

* The *relative* weights of the atoms of the different elements, measured by some standard atomic weight (in this case by hydrogen = 1), are for purposes of simplicity termed the *atomic weights*. This expression will be used in the future. For the same reason the *relative* molecular weights are termed *molecular weights*.

† In some cases the first letter of the Latin name is used; e.g., K for *Kalium* (Latin for Potassium), Na for *Natrium* (Sodium). Where several elements have the same initial a second letter may be added; C = Carbon; Cl = Chlorine; Cu = Copper; Ca = Calcium.

resenting the number of atoms entering into their formation after the respective symbols; i.e., H_2O represents one molecule of water containing two atoms of hydrogen and one of oxygen; HCl, one molecule of hydrogen chloride containing one atom of hydrogen and one of chlorine. NH_3, one molecule of ammonia containing one atom of nitrogen and three of hydrogen. As these symbols represent the uniting atoms, they must of necessity also stand for the atomic weights. For example, the combination H_2O means, not only that two atoms of hydrogen with one atom of oxygen form one molecule of water, but it also represents the fact that two parts of hydrogen are united with 16 of oxygen to produce 18 of water.

Determination of the Number of Atoms united in Molecules by considering the Combining Volumes of Gases. We arrive at the same conclusions regarding the number of atoms uniting to form the molecules of the gaseous compounds which we have studied, if we take into consideration the combining volumes of the elementary gases without regard to their specific gravities. This will be clear from the following reasoning:—

Composition of Hydrogen Chloride. We learned (page 41) that hydrogen chloride is produced by the union of equal volumes of hydrogen and of chlorine. From the principles of the kinetic gas theory, this must mean that equal numbers of molecules of hydrogen and chlorine react to form hydrogen chloride; i.e., one molecule of hydrogen with one molecule of chlorine. We have also learned that the volume of hydrogen chlorine which results is twice as great as the initial volume of hydrogen, so that there must be formed twice as many molecules of hydrogen chloride as there were either of hydrogen or of chlorine. One molecule of hydrogen, which contains two atoms, with one molecule of chlorine, which contains two atoms, therefore, produces two molecules of hydrogen chloride, which contain each one atom of hydrogen and one of chlorine. This result can be expressed by chemical notation as follows:—

$$H_2 \quad + \quad Cl_2 \quad = \quad HCl + HCl.$$

1 mol. of hydrogen + 1 mol. of chlorine = 2 molecules of hydrogen chloride.

1 Vol. Hydrogen	+	1 Vol. Chlorine	=	1 Vol. Hydrogen Chloride	+	1 Vol. Hydrogen Chloride

Now let us suppose that the volume of hydrogen which we selected weighs two grams. From the preceding tables, we know that an equal volume of chlorine must weigh 71 grams. The two grams of hydrogen uniting with 71 grams of chlorine produce 73 grams of hydrogen chloride, which 73 grams occupy twice the volume taken by two grams of hydrogen, and, consequently, a volume of hydrogen chloride equal to that of two grams of hydrogen would weigh 36.5 grams. The specific gravity of hydrogen chloride (hydrogen = 2) is therefore 36.5. Hence the molecular weight of hydrogen chloride is also 36.5, a conclusion which we reached by direct comparison in the first portion of this chapter.

Composition of One Molecule of Water. We learned that water is produced by the union of two volumes of hydrogen and one of oxygen, hence, in its formation, two molecules of hydrogen react with one molecule of oxygen. The volume of water vapor which results is equal to that of the hydrogen with which we started; i.e., it is two volumes, and hence represents two molecules. One molecule of oxygen with two molecules of hydrogen, therefore, produces two molecules of water. This result can be expressed as follows:—

$$H_2 \quad + \quad H_2 \quad + O_2 \quad = H_2O + H_2O.$$

1 mol. of hydrogen + 1 mol. of hydrogen + 1 mol. of oxygen = 2 mols. of water.

1 Vol. Hydro-gen	+	1 Vol. Hydro-gen	+	1 Vol. Oxygen	=	1 Vol. Water Vapor	+	1 Vol. Water Vapor

2 grams hydrogen + 2 grams hydrogen + 32 grams oxygen = 18 grams water + 18 grams water.

If, now, one of these volumes of hydrogen weighs 2 grams, then the same volume of oxygen, as we have seen, weighs 32 grams, and 4 grams of hydrogen with 32 grams of oxygen produce 36 grams of water vapor. This quantity, however, occupies twice the volume taken by 2 grams of hydrogen. Hence the weight of a volume of water vapor equal to that of 2 grams of hydrogen is 18 grams. The specific gravity of water vapor (hydrogen = 2) is consequently 18, and the relative weight of a molecule of water is also 18.

Composition of One Molecule of Ammonia. Three volumes of hydrogen and one of nitrogen unite to form two volumes of ammonia. (See page 94.) Three molecules of hydrogen, therefore, react with one molecule of nitrogen to form two molecules of the compound. As one molecule of nitrogen is capable of forming two molecules of ammonia, it must follow that each nitrogen molecule is divisible into at least two parts, so that each must contain at least two atoms of the element. As each hydrogen molecule also contains two atoms, the formation of ammonia can be represented as follows: —

$$H_2 + H_2 + H_2 + N_2 = NH_3 + NH_3.$$

3 mols. hydrogen + 1 mol. nitrogen = 2 mols. of ammonia.

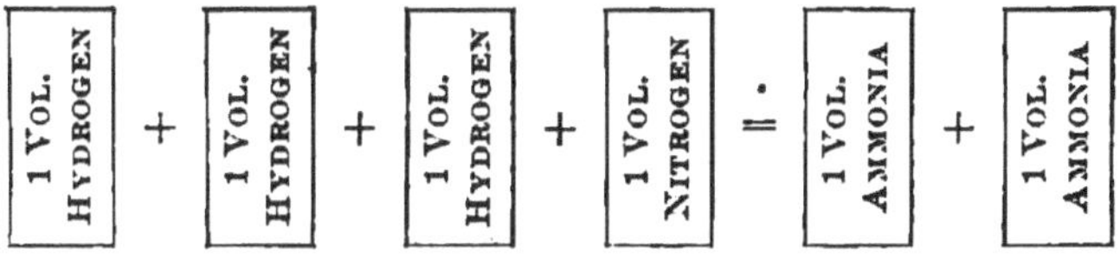

1 Vol. Hydrogen	+	1 Vol. Hydrogen	+	1 Vol. Hydrogen	+	1 Vol. Nitrogen	=	1 Vol. Ammonia	+	1 Vol. Ammonia

If one volume of hydrogen weighs two grams, an equal volume of nitrogen weighs 28 grams. Therefore 6 grams of hydrogen with 28 grams of nitrogen produce 34 grams of ammonia. This quantity of ammonia occupies twice the volume of two grams of hydrogen, hence the weight of a volume of ammonia equal to that of two grams of hydrogen is 17. The specific gravity of ammonia (hydrogen = 2) is therefore 17, and its molecular weight is also 17.

Summary.

1. Each molecule of hydrogen consists of at least two atoms.

2. The weight of a molecule of hydrogen is twice the weight of one atom. The weight of a molecule of hydrogen is the standard for measuring other molecular weights.

3. If hydrogen as 2 is made the standard for measuring the specific gravity of gases, then the specific gravity of any gas is the same as its molecular weight.

4. The determination of the specific gravities, and hence of the molecular weights of gaseous compounds, gives us a means of ascertaining the maximum numbers to be assigned to the relative weights of the atoms of the elements. The minimum numbers for these relative atomic weights are fixed by reasoning that, because we find no gaseous compounds which have less than these minimum quantities of the elements in one molecule, therefore no smaller quantities need be considered.

5. The molecules of the elements, chlorine, oxygen, and nitrogen, consist of at least two atoms.

6. One atom of each of the different elements can be represented by an appropriate symbol. The molecules formed by the union of these atoms can be represented by joining together the symbols, with subscript numbers after each. These numbers indicate the number of atoms of each element which enters into a molecule of the compound.

7. The number of atoms united in the molecule of gaseous compounds can be ascertained by a proper consideration of the combining volumes of the gaseous elements.

CHAPTER XVIII.

THE EXPRESSION OF CHEMICAL CHANGES BY FORMULAE AND EQUATIONS.

THE various considerations advanced in the preceding chapter have shown us that the proper interpretation of the relationships between the combining volumes of gases, and the volumes of the compounds produced by such combinations, determines the molecular weights of those compounds. Knowing the molecular weights, and also the relative parts by weight in which the elements unite, we can, therefore, form a conclusion as to the number of atoms in each molecule, and as to the relative atomic weights. These results, so far as we have gone, can be summed up in the following table:—

FORMULÆ.	MOLECULAR WEIGHTS.	COMPOSITION BY WEIGHT.	ATOMIC WEIGHTS, H = 1.
Hydrogen chloride, HCl	36.5	1 part H, 35.5 Cl	Chlorine, 35.5
Water, H_2O	18.	2 parts H, 16 O	Oxygen, 16.
Ammonia, H_3N	17.	3 parts H, 14 N	Nitrogen, 14.

Composition of the Molecules of Sulphur Dioxide and Sulphur Trioxide. In our previous work we have repeatedly had occasion to mention two other gasifiable compounds (sulphur dioxide and sulphur trioxide). With our present knowledge we are in a position to determine the number of atoms of sulphur and of oxygen which are combined in these two substances.

Sulphur Dioxide. Sulphur dioxide is produced by burning sulphur in oxygen; and if care is taken to determine the relationship between the volume of oxygen used in the combustion, and the volume of sulphur dioxide produced, we shall arrive at the following results: —

When sulphur and oxygen unite to produce sulphur dioxide, the volume of sulphur dioxide formed is the same as the volume of oxygen with which we started. (See page 54.)

As the volume of oxygen and sulphur dioxide are equal, it follows the sulphur dioxide must contain exactly as many molecules as did the oxygen. Let us suppose that a volume of oxygen contains one hundred molecules. Then the sulphur dioxide which would be produced by burning sulphur in this same volume would also contain one hundred molecules. It follows from this that one molecule of sulphur dioxide contains one molecule of oxygen. But as we have seen that one molecule of oxygen contains *two atoms*, it follows that one molecule of sulphur dioxide has two atoms of oxygen. We have decided that the atomic weight of oxygen is 16, provided that of hydrogen be considered as one. Therefore *two* atoms of oxygen, measured in the hydrogen standard, would weigh 32. Since sulphur dioxide has equal parts of sulphur and of oxygen, we can say, in terms of the atomic theory, that *32 parts of sulphur* are joined to *32 parts of oxygen* in sulphur dioxide. Now, it can be decided that 32 represents the weight of one atom of sulphur, hydrogen representing unity; for in no compound of sulphur which is a gas, and the molecular weight of which we consequently know, do we find less than 32 parts by weight of sulphur in one molecule. The specific gravity of sulphur dioxide (hydrogen = 2) is 64. In 64 parts of sulphur dioxide there are 32 parts of sulphur and 32 of oxygen. It follows from this that the atomic weight of sulphur cannot be more than 32, for otherwise we should have to accept the existence of a fraction of an atom of sulphur in one molecule of sulphur dioxide. Therefore, the composition of sulphur dioxide is such that one atom of sulphur is joined to two atoms of oxygen in each molecule. This result can be expressed by the following formula (in which S represents one atom of sulphur) SO_2.

Sulphur Trioxide. When sulphur dioxide is changed into sulphur trioxide, it adds oxygen in such quantity that the resulting compound contains for every one part of sulphur, 1.5 parts by weight of oxygen. Therefore (the 32 parts by weight of sulphur in 64 of sulphur dioxide remaining unchanged), sulphur trioxide contains 32 parts of sulphur to 48 of oxygen (1 : 1.5). Now, 32 represents the relative weight of a sulphur atom, measured in the hydrogen standard, and $48 = (3 \times 16)$ represents the relative weight of three oxygen atoms. Hence a molecule of sulphur trioxide contains, for every atom of sulphur, three atoms of oxygen. The specific gravity of sulphur trioxide in the form of a gas is 80 (hydrogen = 2), so that its molecular weight is also 80 ($H_2 = 2$). In 80 parts of sulphur trioxide there are 32 parts of sulphur and 48 parts of oxygen. A molecule of sulphur trioxide is therefore represented by the formula SO_3.

Formation of Sulphuric Acid from Sulphur Trioxide and Water. Sulphur trioxide unites with water to produce sulphuric acid. (See page 56.) This change is of such a nature that 80 parts of sulphur trioxide take up 18 parts of water to produce 98 parts of pure sulphuric acid. Eighty represents the relative weight of one molecule of sulphur trioxide, and 18 that of one molecule of water. This change can therefore be represented as follows: —

$$H_2O + SO_3 = H_2SO_4 \,*$$

Sulphuric acid, therefore, contains, for every two atoms of hydrogen, one of sulphur, and four of oxygen.

Relative Atomic Weights of Sodium and Potassium. In order to complete our understanding of the chemical changes discussed in the preceding chapters, it is necessary to determine the relative weights of the atoms of potassium and sodium. This is not easily accomplished

* In the member at the right of the sign of equality we have simply added the two terms at the left. The *masses* of matter on both sides are therefore equal.

for gasifiable compounds of the elements in question are not to be obtained by simple means. However, sufficient data are at hand to warrant our forming a conclusion.

An element known as iodine exists which is, in all of its chemical aspects, similar to chlorine. Iodine forms a hydrogen compound, hydrogen iodide, which is a gas like hydrogen chloride. Reasoning similar to that advanced under hydrogen chloride has led us to the conclusion that one molecule of hydrogen iodide contains one atom of hydrogen and one atom of iodine. The molecular weight of hydrogen iodide ($H_2 = 2$) is 127.5. This quantity of hydrogen iodide contains one part of hydrogen and 126.5 parts of iodine. Therefore, the relative weight of one atom of iodine is 126.5, hydrogen being one. Now, potassium reacts with hydrogen iodide just as it does with hydrogen chloride, replacing the hydrogen and producing *potassium iodide.* Potassium iodide, when changed to a gas by great heat, has a specific gravity (hydrogen = 2) of 165.5, hence its molecular weight ($H_2 = 2$) is also 165.5. In 165.5 parts of potassium iodide there are 126.5 of iodine (the proportional part representing one atom of iodine) and 39 parts of potassium. In the absence of any evidence to the contrary, 39 may, therefore, be regarded as the relative weight of one atom of potassium.

Action of Potassium and Sodium on Water in Terms of the Atomic Theory. When potassium acts upon water, one-half the hydrogen in the water attacked is replaced by potassium. An analysis of the product (potassium hydroxide) shows us that 39 parts of the metal take the place of one part of hydrogen. If 39 and one represent the atomic weights of potassium and of hydrogen respectively, then *one atom of potassium replaces one atom of hydrogen* in each molecule of water. This change can be represented as follows: —

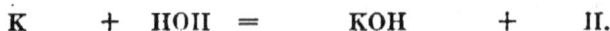

$$K + HOH = KOH + H.$$

1 atom of potassium* + 1 mol. of water = 1 mol. of potassium hydroxide + 1 atom of hydrogen.

Potassium hydroxide, therefore, can be considered as water, in each molecule of which one atom of hydrogen has been replaced

* K represents one atom of potassium, atomic weight 39. This symbol is taken from the Latin *Kalium.*

by one atom of potassium. From the great similarity between potassium hydroxide and sodium hydroxide, it may be presumed that they are alike in their constitution. The latter then may be regarded as water, in each molecule of which one atom of sodium has taken the place of one atom of hydrogen. As 23 parts of sodium replace one part of hydrogen, it may also be presumed that the atomic weight of sodium is 23 if that of hydrogen is one. The action of sodium on water can, therefore, be represented as follows:*—

$$Na + HOH = Na\,OH + H.$$

Neutralization of Hydrochloric Acid in Terms of the Atomic Theory. When sodium hydroxide acts on hydrochloric acid the latter is *neutralized* (see page 48), and the resulting salt (sodium chloride) contains 23 parts of sodium for every 35.5 parts of chlorine. These parts also represent the relative atomic weights of sodium and chlorine, so that sodium chloride is produced by the union of equal numbers of sodium and chlorine atoms, and its formula is represented by Na Cl. The neutralization of sodium hydroxide by hydrogen chloride can now be written in terms of the atomic theory as follows:—

$$Na\,OH + H\,Cl = Na\,Cl + HOH.$$

Adding the atomic weights of the members to the left of the sign of equality, and doing the same with those on the right, we have—

Sodium, atomic weight,	23	Sodium, atomic weight,	23
Oxygen, atomic weight,	16	Chlorine, atomic weight,	35.5
Hydrogen, atomic weight,	1	Sodium chloride, total,	58.5
Sodium hydroxide, total,	40		
Chlorine, atomic weight,	35.5	Oxygen, atomic weight,	16
Hydrogen, atomic weight,	1	Hydrogen, two atoms,	2
Hydrochloric acid, total,	36.5	Water, total,	18

and—

$$40 + 36.5 = 58.5 + 18.$$

According to this equation, therefore, 40 parts of sodium hydroxide will exactly neutralize 36.5 parts of hydrochloric acid, producing 58.5 parts of sodium chloride and 18 parts of water. This

* The symbol Na represents one atom of sodium, atomic weight 23. This symbol is taken from the Latin *Natrium*.

theoretical result is exactly in accordance with what we actually found in the experiments on neutralization (see page 62). No other conclusion is admissible, since the atomic theory to have any stability must of necessity coincide with known facts.

Neutralization of Sulphuric Acid by Sodium and Potassium Hydroxide in Terms of the Atomic Theory. In neutralizing sulphuric acid with potassium hydroxide or sodium hydroxide, we learned that *two* salts of each metal could be produced. These salts were designated as being *primary* and *secondary*. In the primary salt, one-half the hydrogen of the combining sulphuric acid is replaced by the metal (sodium or potassium). In the secondary salt, all the hydrogen is so replaced. In terms of the atomic theory, we can express these facts as follows:—

1. $\begin{cases} NaOH + H_2SO_4 = NaHSO_4 + H_2O. \\ NaOH + NaHSO_4 = Na_2SO_4 + H_2O. \end{cases}$

2. $\begin{cases} KOH + H_2SO_4 = KHSO_4 + H_2O. \\ KOH + KHSO_4 = K_2SO_4 + H_2O. \end{cases}$

Adding the members to the right and those to the left of the sign of equality as we did before, we have —

Sodium, atomic weight,	23	Sodium, atomic weight,	23
Oxygen, atomic weight,	16	Sulphur, atomic weight,	32
Hydrogen, atomic weight,	1	4 atoms of oxygen,	64
Sodium hydroxide, total,	40	Hydrogen, atomic weight,	1
		Primary sod. sulphate, total,	120
Sulphur, atomic weight,	32		
4 atoms of oxygen,	64	Oxygen, atomic weight,	16
2 atoms of hydrogen,	2	2 atoms of hydrogen,	2
Sulphuric acid, total,	98	Water, total,	18

and —

$$40 + 98 = 120 + 18.$$

In forming the secondary sulphate of sodium, we would, therefore, have —

Sodium hydroxide, 80 parts by weight,
Sulphuric acid, 98 parts by weight,
Secondary sodium sulphate, 142 parts by weight,
Two molecules of water, 36 parts by weight,

and —

$$80 + 98 = 142 + 36.$$

So that 40 and 80 parts of sodium hydroxide are capable of reacting with 98 parts of sulphuric acid to produce 120 and 142 parts of the primary and the secondary sulphate of sodium respectively. These results are in accordance with those which we found experimentally on page 63. With potassium hydroxide and sulphuric acid we have —

Potassium, atomic weight,	39	Potassium, atomic weight,	39
Oxygen, atomic weight,	16	Sulphur, atomic weight,	32
Hydrogen, atomic weight,	1	4 atoms of oxygen,	64
Potassium hydroxide, total,	56	Hydrogen, atomic weight,	1
		Potass. pr. sulphate, total,	136
Sulphur, atomic weight,	32		
4 atoms oxygen,	64	Oxygen, atomic weight,	16
2 atoms of hydrogen,	2	2 atoms of hydrogen,	2
Sulphuric acid, total,	98	Water, total,	18

and —

$$56 + 98 = 136 + 18.$$

In forming the secondary sulphate of potassium we should, therefore, have —

Potassium hydroxide, 112 parts,
Sulphuric acid, 98 parts,
Secondary potassium sulphate, 174 parts,
Two molecules of water, 36 parts,

and —

$$112 + 98 = 174 + 36.$$

So that 56 and 112 parts of potassium hydroxide are capable of reacting with 98 parts of sulphuric acid to produce 136 and 174 parts of the primary and secondary sulphate of potassium respectively. These results are in accordance with those which we found experimentally on page 63.

Formation of Ammonium Chloride in Terms of the Atomic Theory. The only remaining compound which was discussed at length in the preceding work is ammonium chloride. This substance is produced by the union of equal volumes of ammonia and hydrogen chloride, therefore it is formed from equal numbers of molecules of ammonia and hydrogen chloride. In other words, one molecule of ammonia with one molecule of hydrogen chloride forms one molecule of ammonium chloride. Under proper conditions ammonium chloride can be vaporized without change. Its specific gravity (hydrogen = 2) is then 53.5, so that its molecular weight is also 53.5. In this 5.35 parts of ammonium chloride there are 14 of nitrogen, 4 of hydrogen, and 35.5 of chlorine. The composition of one molecule of ammonium chloride can, therefore, be represented by the formula NH_4Cl, and its formation from ammonia and hydrogen chloride as follows: —

$$NH_3 + HCl = NH_4Cl.$$

The group of elements NH_4, composed of one atom of nitrogen and four of hydrogen, is the one which under the name of ammonium can take the place of one atom of potassium or sodium in chemical compounds. (See page 99.)

Table of Atomic Weights and Formulæ of Compounds already studied. The atomic weights of the elements which we have encountered, the formulæ of the compounds, and the reactions leading to their formation, are summed up in the following table: —

HYDROGEN ATOMIC WEIGHT = 1.

ELEMENTS.	SYMBOLS.	ATOMIC WEIGHTS.	COMPOUNDS.	FORMULÆ.	COMPOSITION BY WEIGHT.
Chlorine	Cl	35.5	Hydrogen chloride	HCl	1 of hydrogen, 35.5 of chlorine.
Oxygen	O	16	Water	H_2O	2 of hydrogen, 16 of oxygen.
Nitrogen	N	14	Ammonia	H_3N	3 of hydrogen, 14 of nitrogen.
Sodium	Na	23	Sodium hydroxide	$NaOH$	23 of sodium, 16 of oxygen, 1 of hydrogen.
Potassium	K	39	Potassium	KOH	39 of potassium, 16 of oxygen, 1 of hydrogen.
Sulphur	S	32	Sulphur dioxide	SO_2	32 of sulphur, 32 of oxygen.
			Sulphur trioxide	SO_3	32 of sulphur, 48 of oxygen.
			Sulphuric acid	H_2SO_4	32 of sulphur, 64 of oxygen, 2 of hydrogen.
			Ammonium chloride	NH_4Cl	14 of nitrogen, 35.5 of chlorine, 4 of hydrogen.

REACTIONS.

$$Na + HOH = Na\,OH + H.$$

Action of sodium on water, 1 atom of sodium replacing 1 atom hydrogen.

$$K + HOH = KOH + H.$$

Action of potassium on water, 1 atom of potassium replacing 1 atom hydrogen.

$$Na\,OH + H_2SO_4 = Na\,HSO_4 + H_2O.$$

Action of sodium hydroxide on sulphuric acid, 1 atom of sodium replacing 1 atom hydrogen.

$$2\,Na\,OH + H_2SO_4 = Na_2SO_4 + 2\,H_2O.$$

Action of sodium hydroxide on sulphuric acid, 2 atoms of sodium replacing 2 atoms hydrogen.

$$KOH + H_2SO_4 = KHSO_4 + H_2O.$$

Action of potassium hydroxide on sulphuric acid, 1 atom of potassium replacing 1 atom hydrogen.

$$2\,KOH + H_2SO_4 = K_2SO_4 + 2H_2O.$$

Action of potassium hydroxide on sulphuric acid, 2 atoms of potassium replacing 2 atoms hydrogen.

$$NH_3 + H\,Cl = NH_4Cl.$$

Action of ammonia on hydrogen chloride, forming ammonium chloride.

Meaning and Uses of Chemical Equations. The changes taking place during chemical reactions can, as we have seen, be expressed in the form of equations. To the left of the sign of equality are those substances which are about to react with each other, to the right are the new and more stable bodies which result from those reactions. The same elements appear in both. These equations represent what takes place between the smallest particles of the substances taking part in the chemical changes. As what is true of these smallest particles must also be true of the large masses which are composed of them, it follows that these equations also represent the changes which take place in visible portions of matter. As the symbols of the various elements represent the relative atomic weights of these elements, the equations must represent both the relative weights of the substances which are reacting, and the relative weights of the masses handled. In practice there is frequently present an excess of one or the other of the substances undergoing chemical change; but as such an excess does not enter into the chemical formation of the compound produced, it need not be considered in the notation. One serious defect is evident in equations such as we have considered, — they do not take cognizance of the changes of energy during chemical reactions. They simply concern themselves with the reacting masses. Nevertheless, they give us a short and convenient way of illustrating one of the main features of the various chemical changes, and involve a considerable saving in time. Two things must be borne in mind; first, chemical notation and chemical equations are not absolutely essential to an understanding of chemical change; secondly, chemical equations must be used to express only those reactions which have actually been studied. They are merely brief and convenient ways of expressing knowledge already acquired by experiment. The careful study of all the phenomena attending a chemical change must be carried out before the results of such a study can properly be brought in the form of a chemical equation.

Summary.

The changes taking place during chemical reactions can be expressed in the form of equations. These equations do not consider either the excess of one or the other of the reacting bodies, or the changes of energy which take place during reactions. They do, however, express the masses of the bodies undergoing chemical change, and these masses must be in accordance with facts which have been experimentally proven. The reactions which have been encountered in this book are given, in the form of equations, on pages 143, 144, and 147.

Having discussed the atomic theory, the next step in the work is to consider the chemistry of another element, and of its compounds.

CHAPTER XIX.

CARBON.

Carbon is one of the most important chemical elements. In nature it occurs uncombined in three modifications, two of which (diamond and graphite) are of crystalline structure, while the third is not crystalline (coal, charcoal, lamp-black, etc.).

Various Modifications of Carbon. That all these forms are really modifications of the same element is proved by the fact that they all yield the same substance (carbon dioxide) when they are burned in oxygen. By proper means, not-crystalline carbon can be regained from carbon dioxide, no matter whether it has been formed by the combustion of diamond, graphite, or coal.

Diamond. The diamond is an extremely hard substance, very often transparent, with a great power of refracting light, and brilliant lustre when polished. Not-transparent black diamond is also found.* When heated in the air to a bright white heat, diamond burns, forming carbon dioxide, and leaving only a slight trace of ash. The diamond may be classed among the rare minerals.

* Diamonds were originally imported into Europe from the East Indies. From this portion of the world and from Borneo came the only specimens known until the year 1727, when large diamond fields were discovered in Brazil. In 1867 the diamond fields of South Africa were opened.

Graphite. Graphite (also called plumbago or black lead) is the second crystalline form of carbon. Sometimes graphite is found in a pure state, as in some parts of Ceylon, where beds of from twenty to thirty feet in thickness occur. Sometimes it is very impure, being mixed with so much foreign material that it is entirely unfit for use. Graphite may be artificially prepared by the crystallization of carbon from melted iron. It then can be separated in the form of small, delicate scales when the iron is dissolved away by acids. Graphite is used in the manufacture of lead pencils, in making infusible crucibles, and as a lubricator. It is adapted to the latter purpose because it is soft and scaly. It is grayish-black in color, and has almost a metallic lustre. When burned in the air it forms carbon dioxide, and leaves an ash which varies in quantity according to the amount of impurity originally present.

Decomposition of Animal and Vegetable Substances. The compounds formed in animal and vegetable organisms, and classed under the general head of organic substances, are produced by the union of very few elements, namely, carbon, hydrogen, oxygen, nitrogen, sulphur, and phosphorus, although occasionally other elements are found. When exposed to the action of moisture and of the air, organic substances are completely decomposed, changing for the most part into gaseous products (carbon dioxide, ammonium compounds, etc.). Vegetable fibres when protected by a layer of water or of earth, so that an insufficient supply of oxygen is present, decompose very slowly, certain constituents, especially oxygen and hydrogen, pass off in other combinations, and the vegetable matter becomes changed, first into peat, and then into bituminous (soft) coal. At the same time the percentage of carbon increases. Peat, brown coal, bituminous coal, and anthracite coal are successive steps in the process of vegetable decomposition. When the anthracite stage is reached, the changes have become so complete that a black, shiny, homogeneous mass remains, and the original vegetable structure has entirely disappeared, or become so indistinct that special means must be taken for its detection. The pressure to which the dead organic structures are subjected has a material influence on the rapidity with which a peat formation is changed to anthracite. In districts of Russia where there has not been great pressure, a brown coal (lignite), scarcely to be distinguished from peat, is found in places where the age of the deposit would lead one to expect anthracite.[59] *

Formation of Coke. Changes similar to those produced in the transformation of vegetable matter into coal can be brought about by heating the same class of substances in retorts, if the air is excluded. By means of this process, certain portions pass off as gases and liquids,† while a not-crystalline variety of carbon is

* The following table illustrates the increase in the percentage of carbon during the change from wood to anthracite coal. This table takes into consideration only the combustible materials, and not the ash:—

Wood contains 50 per cent of carbon, 6 per cent of hydrogen, and 44 per cent of oxygen and nitrogen.

Peat contains 60 per cent of carbon, 5.7 per cent of hydrogen, and 34.3 per cent of oxygen and nitrogen.

Bituminous coal contains 87 per cent of carbon, 5.6 per cent of hydrogen, and 7.4 per cent of oxygen and nitrogen.

Anthracite contains 94 per cent of carbon, 3.4 per cent of hydrogen, and 2.6 per cent of oxygen and nitrogen.

† Gaseous products,—Illuminating gas, ammonia, sulphuretted hydrogen, etc.

Liquid products,—Water, benzene, toluene, carbolic acid, etc.

Solid products,—Naphthaline, anthracene, etc.

left behind as a black shiny mass termed coke. The commercial production of coke is generally brought about by heating bituminous coal.

Charcoal. Wood charcoal is produced by the imperfect combustion of wood, animal charcoal, by a similar treatment of animal refuse. All forms of charcoal have a remarkably pronounced tendency to absorb coloring matters from solutions.[60]

Lamp-black. The purest form of not-crystalline carbon is lamp-black. This results when carbon and hydrogen compounds are burned in an imperfect supply of oxygen. The lamp-black of commerce is obtained by burning resinous pine wood, tar, or some kinds of bituminous (soft) coal. The substance is collected on coarse cloths hung over the burning wood, which is placed in suitable chambers. Lamp-black is used in the manufacture of India ink and printers' ink.

CHAPTER XX.

CARBON DIOXIDE

CARBON is capable of forming two oxides which are distinguished by the terms *carbon monoxide* and *carbon dioxide*. Of these the latter is the easier to produce and study. A consideration of the former will be postponed to a subsequent chapter.

Formation of Carbon Dioxide and the Formula for one Molecule of Carbon Dioxide. When carbon is burned in the air or in oxygen, it unites with the oxygen to produce a colorless gas which is an oxide of carbon. If the combustion takes place in a closed space, so that the volume of gas present at the beginning of the operation and at the end can be carefully measured, it will be seen that no change in the total volume has occurred.[61] All of the oxygen has been consumed, and carbon dioxide has taken its place. A similar result, it will be remembered, was encountered in studying the formation of sulphur dioxide (see page 54). We are, therefore, justified in drawing similar conclusions as to the structure of the carbon dioxide molecules. We can assume that as there is no change in the total volume of gas, and no change in the total number of molecules, then n molecules of oxygen, as a consequence, produce n molecules of carbon dioxide, and (if $n = 1$) one molecule of oxygen enters into the formation of one molecule of carbon dioxide. Since one molecule of oxygen contains two atoms, it must follow that one molecule of carbon dioxide also contains two atoms of oxygen. The specific gravity of carbon dioxide (hydrogen = 2) is 44. Its molecular weight ($H_2 = 2$) is also 44. In 44 parts of carbon dioxide there are 32 of oxygen and 12 of carbon. Therefore, 12 represents the *maximum* atomic weight of carbon, provided that of hydrogen is one. That 12 is also the minimum atomic weight is probable from the fact that no gasifiable carbon compound contains less than 12 parts of carbon (measured in the hydrogen standard) in one molecule. The 12 parts of carbon in the molecule of carbon dioxide, therefore, represent one atom of carbon, and the 32 parts of oxygen represent two atoms of oxygen, and the formula for carbon dioxide is CO_2. This formula, is, therefore, parallel to that for sulphur dioxide:—

CO_2	SO_2
Carbon dioxide,	Sulphur dioxide.

Preparation of Carbon Dioxide. If an iron tube in which are placed pieces of charcoal be heated in a furnace, and pure oxygen be then passed over the hot carbon, the carbon dioxide which is formed can be collected in bottles. This gas has a higher specific gravity than the atmosphere, and consequently will flow downward, just as a liquid would.[62]

Properties of Carbon Dioxide. Carbon dioxide is a colorless gas with a specific gravity (air = 1) of 1.529 or ($H_2 = 2$) of 44. The latter number is also its molecular weight. Because carbon dioxide has such a high specific gravity, it can be poured downward from any vessel containing it. For this reason it collects at the bottoms of wells or mines into which the gas is escaping. Carbon dioxide does not burn in oxygen, neither does it give up any portion of its oxygen readily enough to support combustion. Cold and pressure combined can condense it to a liquid which boils at

− 78.2°. Its vapor tension (see page 70) at 0° is 36 atmospheres, its critical temperature* is 39.9°, and its vapor pressure at this point is 73 atmospheres. When carbon dioxide is rapidly evaporated in a vacuum, its temperature sinks to − 97°, and it then changes to a snow-like solid. Any considerable increase in the quantity of carbon dioxide normally present in the air (3 parts in 10,000, see page 84) causes a marked feeling of discomfort to those breathing it, and a proportion of approximately 10 per cent will cause death. The reason for this is that the pressure of carbon dioxide in the inhaled air is equal to that of the carbon dioxide passing off from the lungs. When this condition is reached, no elimination of carbon dioxide can take place, the normal processes of life are interfered with, and the animal dies. Plants, when placed in the sunlight, are, on the other hand, capable of absorbing carbon dioxide, changing it into new compounds of carbon (which form a portion of their tissues), and eliminating a portion of the oxygen. One cubic centimetre of water can dissolve about its own volume of carbon dioxide at ordinary temperatures.

Formation of Carbonates from Carbon Dioxide and Bases. If carbon dioxide is passed into a solution of potassium hydroxide, complete absorption takes place. If this operation is continued until the base has taken up all the gas which it possibly can, and if the excess of water used for solution is then evaporated, there remains a salt-like body which differs entirely from potassium hydroxide. If a little of this salt is placed in a test-tube, and some solution of hydrochloric acid is added, a violent evolution of gas will take place. This gas will neither burn nor support combustion, and in all its properties is identical with carbon dioxide. We find, therefore, that carbon dioxide can unite directly with potassium hydroxide to produce a new body. By carefully measuring the quantity of carbon dioxide obtained from a given weight of this body by adding hydrochloric acid, we learn that exactly 44 parts of carbon dioxide (representing one molecule) have united with 56 parts of potassium hydroxide (represented by the formula

* I.e., the temperature above which no pressure can convert it to a liquid.

KOH, see page 142). The reaction which takes place can be represented as follows:—

KOH	+	CO_2	=	$KHCO_3$
potassium hydroxide	+	carbon dioxide	=	The product of addition.
56 parts	+	44 parts	=	100 parts.

If, now, two grams of this salt are placed in a test-tube of infusible glass, and then gently heated to below red heat, a gas will pass off which can be shown to have the properties of carbon dioxide. At the same time water is formed, as can be seen by the drops of that liquid collected on the sides of the tube in which the salt is heated. After no more carbon dioxide passes off, the water can be removed by placing the tube in an air-bath heated to about 150°. After cooling, the two grams of the salt will be found to have lost .62 grams of carbon dioxide and water, while 1.38 grams of a second salt-like body will remain. This body, on addition of hydrochloric acid, will give off a gas, which is carbon dioxide. In this respect it resembles the first salt which was formed. The change taking place on heating can therefore be represented as follows:

$KHCO_3$ =	K_2CO_3	+	H_2O	+	CO_2
2.00 parts =	1.38 parts of the second salt	+	.18 parts of water	+	.44 parts of carbon dioxide. 63

The Primary and Secondary Carbonates of Potassium are Salts of an Acid, H_2CO_3. The two salt-like bodies produced, first by the addition of carbon dioxide to potassium hydroxide, and second by heating the substance so formed, can be considered as derived from an acid to which we assign a formula, H_2CO_3. This acid does not in reality exist. Whenever any attempts have been made to isolate it as a chemical individual, it has at once broken down into water and carbon dioxide. The salts produced by substituting metals for the hydrogen of this acid are, however, as we have seen, permanent. This acid is termed *carbonic acid*, and the salts derived from it by substituting metals for the hydrogen are called *carbonates*. The two bodies just examined are, therefore, *primary* and *secondary potassium carbonates*. The former is produced by replacing one-half of the hydrogen in a given quantity of carbonic acid by the metal potassium, and the latter by replacing all of the hydrogen in the same way. The learner will recall in the discussion of a similar case the *primary* and *secondary* sulphates (see pages 60 and 61), and the formation of the two carbonates of potassium can be compared to the neutralization of sulphuric acid by potassium hydroxide. This relationship will be made clearer if we compare the formulæ assigned to the sulphates of potassium with those just given for the carbonates:—

$KHSO_4$, primary sulphate of potassium.
K_2SO_4, secondary sulphate of potassium.
$KHCO_3$, primary carbonate of potassium.
K_2CO_3, secondary carbonate of potassium.

Difference between the Primary and Secondary Salts of Potassium. The *primary* potassium salts are those in which one part of hydrogen in the acids has been replaced by 39 parts of potassium, and the *secondary* salts are those in which two parts of hydrogen have been replaced by 78 (2×39) parts of potassium. The primary sulphate of potassium is converted into the secondary sulphate (page 59) by the addition of potassium hydroxide. If the relationship between the carbonates and the sulphates is to hold good, it must therefore follow that the primary carbonates can be converted into the secondary by similar means. That this is the case can easily be shown by dissolving one gram of the primary carbonate in a little water, and then adding .56 grams *

* This quantity, in centigrams, represents the sum of the atomic weights of potassium, oxygen, and hydrogen, which go to form potassium hydroxide (see page 145).

potassium, atomic weight,	39
oxygen, atomic weight,	16
hydrogen, atomic weight,	1
Total,	56

of potassium hydroxide.* On evaporating the excess of water from the solution, a salt will remain, which, on addition of hydrochloric acid, gives off carbon dioxide, but which does not evolve that gas on heating. In short, it has all of the properties of the *secondary carbonate.*

The above change can, therefore, be represented in chemical formulæ as follows:—

KOH	+	$KHCO_3$	=	K_2CO_3	+ H_2O
Potassium hydroxide	+	Primary potassium carbonate	=	Secondary potassium carbonate	+ Water.
56 parts	+	100 parts	=	138 parts	+ 18 parts.

The primary carbonate of potassium can, therefore, be converted into the secondary by the addition of potassium hydroxide, just as the primary sulphate is converted into the secondary by the same means.[64]

Occurrence of the Carbonate of Potassium. The secondary carbonate of potassium occurs as one of the constituents of the ashes of land plants, from which it can be extracted by means of water. As this operation was usually performed in iron pots or kettles, the name *potash* was given to potassium carbonate; and since potassium hydroxide is prepared from the latter, the term *caustic potash* was brought into use to designate the hydroxide.

Formation of the Primary and Secondary Carbonates of Sodium. The sodium carbonates are chemically closely related to those of potassium. If, then, in the above reactions, we had made use of sodium hydroxide, instead of potassium hydroxide, the result would have been identical; except that the primary and secondary carbonates of sodium would have taken the place of those of potassium. The chemical formulæ of these two substances would therefore be $NaHCO_3$ and Na_2CO_3, and they would be produced as follows:—

* This should be in the form of a solution where one cubic centimetre contains .056 grams of potassium hydroxide. The solution can be measured in a burette (see page 49, and Experiment 27 of the Appendix).

$$NaOH^{*} + CO_2 = NaHCO_3.$$

Sodium hydroxide + Carbon dioxide = Primary sodium Carbonate.
40 parts + 44 parts = 84 parts.

$$NaHCO_3 + NaOH = NaCO_3 + H_2O.$$

Primary sodium carbonate + Sodium hydroxide = Secondary sodium carbonate + Water.
84 parts + 40 parts = 124 parts + 18 parts.

Sodium carbonate is commercially more important than potassium carbonate, as it is used extensively in the arts, and in the preparation of soap and of glass.

Occurrences of Calcium Carbonate, Magnesium Carbonate, and Carbonate of Iron. The carbonates of several other metals are extensive and valuable mineral constituents of the earth's crust. The carbonate of calcium occurs in a massive, not-crystalline form, as *limestone*, in blocks formed of minute, not-separable crystals, as *marble*, and in well-formed, often transparent crystals, as *calcite*, peculiarly developed forms of which are termed *Iceland spar*. The shells of mollusks, and the protective coating of many infusoria, consist largely of calcium carbonate.

Magnesium carbonate, as the mineral magnesite, occurs in extensive deposits. Magnesium and calcium carbonates (*dolomite*) is the chief mineral in a range of mountains in the Alps. The carbonate of iron is valuable as an ore.

Preparation of Carbon Dioxide by the Action of Acids on Carbonates. All of the carbonates are more or less readily decomposed by acids. The result in each case is the production of the salt of the acid which is used with the metal contained in the carbonate. At the same time carbon dioxide and water are formed. We found this to be the case in studying potassium carbonate, and an examination of other carbonates will show the same result. For example, calcium carbonate with

* NaOH represents: —

One atom of sodium, atomic weight,	23
One atom of oxygen, atomic weight,	16
One atom of hydrogen, atomic weight,	1
Total,	40

(See page 143.)

hydrochloric acid, forms calcium chloride, carbon dioxide, and water.

Calcium carbonate, in the form of marble, can be easily handled in the laboratory. It is so readily attacked by hydrochloric acid that the reaction just given is the one universally employed in preparing carbon dioxide for laboratory use.[65]

Most Carbonates are Insoluble in Water. Formation of the Soluble Primary Carbonate of Calcium from the Insoluble Secondary. The carbonates of most metals do not dissolve in water. In our present work we may consider that only the carbonates of potassium and of sodium are soluble. As an illustration we can show that if carbon dioxide is passed into a solution of *calcium hydroxide* (lime-water), the insoluble *carbonate of calcium* separates as a white powder. If, however, gas is added after the separation of the calcium carbonate is completed, the latter substance will gradually dissolve in the excess of carbon dioxide solution, because the primary carbonate of calcium (which is soluble in water) is produced by this means. When the solution so formed is boiled, carbon dioxide passes off, and the insoluble *secondary carbonate of calcium* is again produced. This happens because *primary calcium carbonate* is less stable than primary potassium carbonate, and hence it decomposes at a lower temperature. This change is parallel to the decomposition of primary potassium carbonate by heat, which we studied at the beginning of this chapter.[66]

Occurrence of the Primary Calcium Carbonate in Spring and River Water. Owing to the action of carbon dioxide on the carbonate of calcium, which, in the form of mineral deposits, constitutes a large portion of the beds of springs and rivers, part of the limestone is dissolved as primary calcium carbonate. Water, which contains this substance in solution, will, when boiled, deposit secondary calcium carbonate on the walls of the kettle. This phenomenon is always observed where hard water is boiled. In addition to the primary calcium carbonate, calcium sulphate, which is also slightly soluble in water, is generally found in spring and river water. This substance is not separated by boiling. Waters which contain either primary calcium carbonate, or calcium sulphate, or both, are termed *hard waters.* Those which have only primary calcium carbonate are temporary hard waters, for this constituent can be removed by boiling; those which contain calcium sulphate are permanent hard waters.

Summary.

1. When carbon is burned in the air, carbon dioxide is produced.

2. Carbon dioxide, in each molecule, contains one atom of carbon and two of oxygen. Its formula is CO_2.

3. Carbon dioxide is absorbed by potassium hydroxide. When a solution of potassium hydroxide is completely saturated by carbon dioxide, it contains primary potassium carbonate, the formula of which is $KHCO_3$.

4. When the primary carbonate of potassium is heated, or when potassium hydroxide is added to it, it is changed into the secondary carbonate of potassium, the formula of which is K_2CO_3. There is also a primary and a secondary carbonate of sodium to which respectively the formulæ $NaHCO_3$ and Na_2CO_3 can be assigned.

5. The carbonates of a number of other metals (calcium, magnesium, iron) form extensive mineral deposits. These carbonates are insoluble in water, while those of potassium and sodium are soluble. The carbonates, on addition of acids, liberate carbon dioxide.

6. The primary carbonates are all soluble in water. The above insoluble minerals are, therefore, partly brought into solution by the combined action of water and the dissolved carbon dioxide.

7. When the water containing them is boiled, these primary carbonates are decomposed, and the insoluble secondary carbonates are deposited. From primary calcium carbonate in temporary hard water, a deposit of secondary calcium carbonate is produced.

CHAPTER XXI.

CARBON MONOXIDE AND METHANE (HYDROGEN CARBIDE).

THE second oxide of carbon is carbon monoxide. This gas is easily produced by taking away a portion of the oxygen from carbon dioxide.

Preparation and Properties of Carbon Monoxide. If carbon dioxide is passed through an iron tube containing pieces of charcoal heated to redness, and then into bottles filled with water and inverted over a vessel filled with the same liquid, the gas collected will differ from carbon dioxide. It will not be absorbed by the potassium hydroxide as carbon dioxide is; it will burn in the air when a lighted taper is applied to the mouth of the jar containing it; and it is scarcely soluble in water. This new gas is termed *carbon monoxide.* It is intensely poisonous. Its specific gravity (hydrogen = 2) is 28, its molecular weight ($H_2 = 2$) is also 28. In 28 parts there are 12 of carbon and 16 of oxygen. These numbers, as we have seen, represent the relative atomic weights of carbon and oxygen respectively, so that one molecule of carbon monoxide contains one atom of carbon and one of oxygen. Its structure can therefore be represented by the formula CO.[67]

Reduction. The process of taking away oxygen from carbon dioxide to form carbon monoxide is the reverse of the oxidation by which carbon monoxide is converted into carbon dioxide. Such a process, in chemical language, is termed *reduction.*

Change in Energy when Carbon Monoxide burns to form Carbon Dioxide. When carbon monoxide burns, a large amount of heat is given off. The carbon dioxide produced, therefore, possesses less chemical energy than the monoxide. We can determine the total amount of heat evolved in burning a given weight of carbon. We can also ascertain the total heat produced when carbon monoxide, containing the same weight of carbon, is converted into carbon dioxide. The difference between these two quantities must necessarily give us the heat given off in forming the equivalent amount of carbon monoxide.* By this means it has been ascertained that carbon, in being partially oxidized to carbon monoxide, gives off kinetic energy, and that the resulting gas contains less chemical energy than did its original constituents, carbon and oxygen. Furthermore, carbon monoxide, in burning to carbon dioxide, again gives off energy. The energy changes attending the formation of carbon dioxide can, therefore, be considered as taking place in two stages; but the total amount of kinetic energy evolved when a given weight of carbon burns to carbon dioxide is the same as if the same amount were first changed to carbon monoxide and then to carbon dioxide. The total kinetic energy manifested during a chemical change is, therefore, independent of the intermediary bodies which may be formed. This is exactly the case with a weight which is allowed to fall from a certain height to the ground. The total energy given off is obviously the same, no matter whether it be allowed to descend without interruptions, or whether it be stopped at intervals. It will obviously require the same amount of energy to bring it back to its original position, no matter what were the circumstances attending its fall.

* I.e., the amount of carbon monoxide containing the above given weight of carbon.

Far more important than carbon monoxide, for the purposes of our present work, is the hydrogen compound of carbon which is known as *methane*, or hydrogen carbide.

Methane; Natural Occurrence and Formation. Hydrogen and carbon cannot be brought to unite directly except with the greatest difficulty, and then only in the smallest quantity. In this respect carbon resembles nitrogen (see page 88). It is not difficult, however, to produce methane from other compounds which contain both carbon and hydrogen. For example, it is constantly formed at the bottom of swamps and stagnant pools where vegetable matter (largely made up of the two elements in question) is decaying under the influence of micro-organisms, and where the amount of oxygen required for a complete decomposition is not accessible. The methane produced by these processes of decay passes off in the form of bubbles, the number of which can be increased by stirring the bottom of the pool with a stick. It is owing to the above origin that methane is frequently termed "marsh gas." Natural marsh gas is contaminated with carbon dioxide and nitrogen. The "natural gas," which escapes freely from borings in many localities, consists largely of methane, which probably found its origin in the decomposition of vegetable matter buried in past geologic eras. The remains of this now constitute bituminous and anthracite coal. Owing to similar reasons, methane frequently occurs in coal-mines.

Preparation of Methane for Laboratory Purposes. In preparing methane for laboratory purposes, we have to make use of a compound known as *sodium acetate*, with the exact constitution of which we need not at present be acquainted. It is sufficient for our purposes to know that it is formed of carbon, hydrogen, oxygen, and sodium, and that, when heated in the presence of a mixture of sodium hydroxide and calcium oxide, it separates, in the form of *methane*, a portion of the carbon and all of the hydrogen which it contains.[68]

Properties of Methane. Methane is a gas, colorless, odorless, and tasteless. If a jar of it is collected over water, and a lighted taper is applied to the jar, which has been removed from the water and held mouth downward, we shall find that the gas burns with a nearly non-luminous flame. This flame greatly resembles that of hydrogen, but, nevertheless, can be distinguished from it because the hydrogen flame is entirely non-luminous. The specific gravity of methane is less than that of air. This can easily be shown by filling two cylinders with methane, and holding one mouth upward and the other mouth downward for a few seconds. The former will no longer contain a combustible gas, for its methane will have risen into the air. The latter will have retained all its methane, and if a lighted taper is applied, will show the characteristic flame of that substance. Methane is nearly insoluble in water.[69]

Determination of the Chemical Formula of Methane. In determining the chemical formula of methane, we cannot have so easy a task as we did when engaged in similar studies with hydrogen chloride, water, or ammonia. Both constituents entering into the formation of each of the latter compounds are gases, and for that reason a study of the uniting gas volumes together with the fundamental theory that in equal volumes of gases there are equal numbers of molecules, directly led us to the structure of those three substances, and thence to the formulæ, HCl, H_2O, and NH_3. In methane, however, one of the constituent elements (carbon) is a solid under any conditions which we can command. Therefore, any view which may be held as to the relative volume of that element entering into the formation of the gas cannot be based upon direct measurement. It must be established both by reasoning from our experience with the other hydrogen compounds (hydrochloric acid, water, and ammonia), and by noting chemical changes to which we can subject methane.

Previous experience has shown us that we have three kinds of gaseous hydrogen compounds of the not-metals which, we may say, belong to three distinct types. The first of these, hydrogen chloride, is produced by the union of one volume of hydrogen with one volume of chlorine, the second by the union of two volumes of hydrogen with one volume of oxygen, and the third by the union of three volumes of hydrogen with one volume of nitrogen. In each of these cases one volume of the not-metal is united with one or more volumes of hydrogen. If methane is a compound which is at all parallel, it seems reasonable to suppose that it contains one volume of carbon united with one or more volumes of hydrogen. This volume of hydrogen is consequently to be ascertained by chemical changes to which we can subject methane.

Volumetric Composition of Methane by Explosion with Oxygen. When methane burns in oxygen it produces carbon dioxide and water. This fact gives us a means of ascertaining the volume of hydrogen which is contained in one volume of methane.

Introduce about ten cubic centimetres of methane into a eudiometer tube over mercury, and then admit omewhat more than twice this volume of pure oxygen. Heat the eudiometer to the temperature of boiling water by means of a steam-jacket (see page 26, and Experiment 11 of Appendix), and explode the gas mixture by an electric spark, exactly as was done in the study of the combining volumes of oxygen and hydrogen. If all due precautions have been taken, it will be found that *no change in the total volume* takes place when methane and oxygen are converted into carbon dioxide and water. There are consequently as many molecules of carbon dioxide and water produced as there were molecules of methane and oxygen before the explosion. If, now, the whole apparatus is allowed to cool, the *water vapor* will be changed to a liquid, so that its volume can be neglected in the subsequent calculation. As the water condenses, the mercury will rise, owing to the contraction of the total gas volume, and, when its position is constant, there will have been exactly *twenty* cubic centimetres diminution. These twenty cubic centimetres must therefore represent the *volume of water vapor* which was formed from ten cubic centimetres of methane.* If a very small piece of caustic potash is now introduced at the bottom of the tube, it will rise to the top of the mercury, and absorb the carbon dioxide which was produced by the explosion, so that a further contraction will take place. This latter will be *exactly* ten cubic centimetres, which is equal to the volume of methane originally present. The gas which finally remains is oxygen, and represents the excess of that substance which was added. The results of the above measurements can be tabulated as follows:—

1. Volume of methane = 10 cubic centimetres.
2. Volume of gases at 100° before and after the explosion, unaltered.
3. Contraction, owing to condensation of water vapor, 20 cubic centimetres.
4. Contraction, owing to the absorption of carbon dioxide, 10 cubic centimetres.

Volume of methane = 10 cubic centimetres.
Volume of water vapor = 20 cubic centimetres.
Volume of carbon dioxide = 10 cubic centimetres.

As a final result, therefore, we find the following—

* Of course the above description presupposes that all the volumes mentioned have been calculated to 0° and 760 mm. pressure. In actual practice the height of the barometer, the temperature, the height of the column of mercury in the eudiometer tube, and the volume of gas in the tube, must be noted at the following stages of the experiment: 1. After admitting the methane. 2. After admitting the oxygen. 3. After exploding the gas mixture (this volume is at 100°). 4. After allowing the products of the explosion to cool. In this instance the tension of water vapor must be taken into account, as at this point the water produced by the explosion is present as such. 5. After introducing the piece of caustic potash. The calculation for the tension of water vapor here disappears, for the caustic potash will absorb the water which is present. See pages 68, 69, and 70.

Results of the Explosion of Methane with Oxygen. *One volume of methane, when exploded with oxygen, produces one volume of carbon dioxide and two volumes of water vapor.* As in equal volumes of gases there are equal numbers of molecules, this result means that *one molecule of methane produces one molecule of carbon dioxide and two molecules of water.* Now we have decided that two molecules of water must contain four atoms of hydrogen; and since these four atoms of hydrogen have been furnished by one molecule of methane, it must follow that —

One molecule of methane contains four atoms of hydrogen.

We have also come to the conclusion that *one* molecule of carbon dioxide contains *one* atom of carbon (see page 154); and, since this one atom of carbon must have been furnished by one molecule of methane, it follows that —

One molecule of methane contains one atom of carbon.

Combining these two results we find that:—

One molecule of methane is composed of one atom of carbon and four atoms of hydrogen. Therefore its formula is represented by CH_4.[70]

Formula of Methane from a Consideration of its Specific Gravity. This conclusion as to the formula of methane is borne out by its specific gravity, which is 16 (hydrogen = 2). The molecular weight of methane ($H_2 = 2$) is consequently 16; and in 16 parts of methane we find, on analysis, twelve of carbon and four of hydrogen. In working with carbon dioxide we decided that 12 represented the relative weight of one atom of carbon measured by the standard of one atom of hydrogen.

For the same reason 4 represents four times the weight of one atom of hydrogen; so that, judging from the specific gravity of methane, its structure is such as would be produced by the union of one atom of carbon with four of hydrogen; thus its formula is CH_4.

Summary.

1. Carbon monoxide can be produced by the reduction of carbon dioxide.
2. When carbon monoxide burns, carbon dioxide is produced, heat is given off. Therefore carbon dioxide possesses less chemical energy that carbon monoxide.
3. The energy changes attending the formation of carbon dioxide can be considered as taking place in two stages, the first corresponding to the formation of carbon monoxide, and the second to the conversion of carbon monoxide into carbon dioxide.
4. The total kinetic energy manifested during the conversion of a given weight of carbon into carbon dioxide is independent of any intermediary formation of carbon monoxide.
5. The hydrogen compound of carbon, which is termed methane, is produced in nature by the decomposition of vegetable matter protected from the direct action of the air.
6. The chemical formula of methane cannot be directly determined from combining gas volumes, first, because carbon is not a gas, and second, because the elements carbon and hydrogen do not easily unite directly.
7. The formula of methane can be ascertained by exploding the gas with oxygen and then measuring the volume of water vapor and of carbon dioxide which is produced by this means.

8. One volume of methane, under these circumstances, produces one volume of carbon dioxide and two volumes of water vapor. Hence, one molecule of methane produces one molecule of carbon dioxide and two molecules of water.

9. One molecule of carbon dioxide contains one atom of carbon, and two molecules of water contain four atoms of hydrogen. Hence, methane contains one atom of carbon to every four atoms of hydrogen.

10. Methane contains one atom of carbon and four atoms of hydrogen in each molecule, for its specific gravity (hydrogen = 2) is 16. Hence its molecular weight is 16. Such a molecular weight is equal to the sum of the atomic weights of one atom of carbon (12), and four atoms of hydrogen (4). Therefore the formula for one molecule of methane is CH_4.

CHAPTER XXII.

SUBSTITUTION OF HYDROGEN IN METHANE BY CHLORINE.

Substitution of Hydrogen by Metals. We have learned in discussing the hydrogen compounds which preceded methane, that certain metals could replace a portion, or all, of the hydrogen in those compounds. This process, which was general with all but methane, we termed *substitution*. We also learned that the same kind of substitution takes place when metals act on acids. Therefore salts were considered as being the acids in which a portion, or all, of the hydrogen had been replaced by other metals. If we have carefully observed all the phenomena attending the substitution of hydrogen in hydrogen chloride, water, and ammonia by metals, we shall have noticed that the ease with which the hydrogen is so replaced, and the number of metals capable of entering into such substitutions, diminishes as we pass from hydrogen chloride to water, and from water to ammonia. Finally, in methane we have a hydrogen compound which, under ordinary circumstances, is entirely indifferent to metals. Sodium, for example, can be left in contact with pure, dry methane for any length of time without suffering the slightest alteration.

Difference between the Action of Chlorine and of the Metals. On the other hand, we have a radically different behavior toward chlorine. Chlorine, of course, has no effect on hydrogen chloride; but when we dissolved chlorine in water, and exposed the solution to the sunlight, we saw that hydrochloric acid was formed and oxygen was liberated (see page 44), while with ammonia and chlorine we produced hydrogen chloride and nitrogen (see page 92). In each instance, then, chlorine liberated the not-metal which was present in the hydrogen compound, while it itself appropriated the hydrogen. Obviously these reactions could not be compared with substitution. It is now in order to study the action of chlorine on methane.

Action of Chlorine on Methane. Two glass jars of equal size are taken, one filled with chlorine, the other with methane. A glass cover is placed on each so that no gas can escape. The mouths are then brought together, the glass coverings are removed, and the chlorine and methane are intimately mixed by inverting the jars once or twice, while their mouths are held tightly together.*

If, now, a lighted taper is applied to one of the vessels, the mixture of gases will burn, dense fumes of hydrogen chloride will be given off, while carbon will separate from the methane in the form of soot. Under the circumstances which have been outlined, methane acts toward chlorine just as water and ammonia do,—carbon is separated, while the chlorine appropriates the hydrogen to form hydrogen chloride. An entirely different result, however, is obtained if chlorine and methane are allowed to act upon each other gradually.[71]

For this purpose the apparatus can be used which was employed to demonstrate that equal volumes of ammonia and hydrogen chloride unite to form ammonium chloride (see Experiment 55 of the Appendix).

* This operation must be conducted in a dimly lighted room, and not in the sunlight.

One of the two equal tubes is carefully filled with pure chlorine, the other with methane, and the tips are sealed. The apparatus is placed in strong daylight, and the two gases are allowed to mingle slowly. After some days the color of the chlorine will entirely disappear. If, now, one of the tips is broken off under a solution of blue litmus, the liquid will rise in the apparatus until exactly one-half is filled, while the litmus will turn from blue to red, showing that an acid is present. This acid can easily be shown to be hydrochloric acid.* If the upper tip of the tube is now broken off, and the whole lowered into the litmus solution so as to produce a slight pressure, the escaping gas can be ignited. It will burn with a *greenish* flame entirely different from that of pure methane. This flame is characteristic of burning compounds of carbon, hydrogen, and chlorine. The explanation of the above experiment is simple. The volume of methane reacted with an equal volume of chlorine to produce one volume of the new chlorinated gas. The hydrogen chloride which was formed at the same time, and which is readily soluble in water, was absorbed by the litmus solution. It is, therefore, evident that *one* volume of methane with *one* volume of chlorine produced *one* volume of the new gas and *one* volume of hydrogen chloride. Therefore, one molecule of methane with one molecule of chlorine formed one molecule of hydrogen chloride and one molecule of chlorinated methane. As one molecule of chlorine contains two atoms of that element, the change can be represented as follows:—[72]

$$CH_4 \quad + \quad Cl\,Cl \quad = \quad CH_3Cl \quad + \quad H\,Cl.$$

One mol. of methane + One mol. of chlorine = One mol. of new gas + One mol. of hydrogen chloride.

The First Product of the Action of Chlorine on Methane is Methyl Chloride. The new gas which is produced is called *methyl chloride.* It is *methane* in each molecule of which one atom of hydrogen has been replaced (substituted) by one atom of chlorine. It is evident, therefore, that if chlorine is allowed to act slowly on methane,

* By adding a solution of silver nitrate to the litmus solution, the insoluble chloride of silver will be produced, thus indicating the presence of hydrochloric acid.

substitution takes place, and that the chlorine which is introduced takes the place of the hydrogen which has been expelled.

Chlorination of Methyl Chloride. The operation of substitution can be carried still farther if, when the hydrogen chloride has been removed by means of the litmus solution, one arm of the tube containing methyl chloride (after the action of chlorine on methane) is not opened. Close the stopcock separating the two halves of the apparatus, and drain off the water contained in the lower half. Fill this again with chlorine and seal it, allowing the gases to mix as in the first instance. The phenomena previously observed will be repeated; i.e., the color of the chlorine will disappear, and, when the tube is opened under a litmus solution, it will be found that one volume of hydrogen chloride and one volume of a chlorinated methhl chloride will have been formed. From this it follows that one molecule (volume) methyl chloride with one molecule (volume) of chlorine produces one molecule (volume) of hydrogen chloride and one molecule (volume) of chlorinated methane. This second change can be represented as follows: —[73]

$$CH_3Cl + ClCl = CH_2Cl_2 + HCl.$$

1 mol. methyl chloride + 1 mol. chlorine = 1 mol. new gas + 1 mol. hydrogen chloride.

The Second Product of the Action of Chlorine on Methane is Methylen Chloride. The second chlorinated gas which is produced is called *methylen chloride.* It is methane in each molecule of which two atoms of hydrogen have been replaced (substituted) by two atoms of chlorine.

Chlorination of Methylen Chloride. Methylen chloride is also capable of reacting with chlorine, but in this instance we cannot easily measure the gas volume produced. The reason is that the third substitution product of methane is a liquid at ordinary temperatures, so that on opening the tube under water, the latter will rush in and entirely fill the apparatus, except the very small space occupied by the substitution product. We will soon see, however, that by a careful study of the specific gravity of the vapor of the liquid formed, we can ascertain that this substitution has proceeded like the others, and that —

One molecule of methylen chloride, with one molecule of chlorine, has produced one molecule of the new substitution product, together with one molecule of hydrogen chloride. This change can be represented as follows:—[74]

$$CH_2Cl_2 \quad + \quad Cl\,Cl \quad = \quad CH\,Cl_3 \quad + \quad H\,Cl.$$

1 mol. methylen chloride + 1 mol. chlorine = 1 mol. of the new product + 1 mol. hydrogen chloride

The Third Product of the Action of Chlorine on Methane is Methin Chloride, and the Fourth, Carbon Tetrachloride. The third substitution product of methane is *methin chloride*, commonly known as chloroform. If chloroform is subjected to the continued action of chlorine, the last remaining hydrogen atom in each molecule will be substituted, and we shall finally have a methane in which all the hydrogen has been replaced by chlorine. This last methane is called carbon tetrachloride. It is a liquid at ordinary temperatures, but can easily be converted into a gas.

Specific Gravities and Boiling-Points of Methane and its Substitution Products. It is most instructive to compare the specific gravities of methane and of its chlorine substitution products, when all five substances have been heated to a temperature sufficient to convert them into gases. That this is easily accomplished will be seen from a table giving their boiling-points.

Methane boils at	− 164°.
Methyl chloride boils at	− 23 .7°.
Methylen chloride boils at	+ 40°.
Methin chloride (chloroform) boils at	+ 61°.
Carbon tetrachloride boils at . . .	+ 76°.

Since all these compounds are gases at temperatures below the boiling-point of water (100°) their specific gravities as vapors can easily be ascertained. These specific gravities are given in the following table:—

Specific gravities of methane and its substitution products as gases.

Hydrogen = 2

Methane, specific gravity 16, molecular weight 16, contains 12 parts of carbon and 4 of hydrogen.

Methyl chloride, specific gravity 50.5, molecular weight 50.5, contains 12 parts of carbon, 3 of hydrogen, and 35.5 of chlorine.

Methylen chloride, specific gravity 85, molecular weight 85, contains 12 parts of carbon, 2 of hydrogen, and 71 of chlorine.

Methin chloride, specific gravity 119.5, molecular weight 119.5, contains 12 parts of carbon, 1 of hydrogen, and 106.5 of chlorine.

Carbon tetrachloride, specific gravity 154, molecular weight 154, contains 12 parts of carbon and 142 parts of chlorine.

Atomic Weights of Carbon and Chlorine as ascertained from the Specific Gravities of Methane and the Chlorinated Methanes. The figures in this table furnish additional proof that the weight of an atom of carbon is twelve times that of an atom of hydrogen, since we have here *five* compounds, each containing 12 parts of carbon in every molecule. The atomic weight of carbon cannot, therefore, be more than 12. That it is not less than this number is rendered more probable the more gaseous compounds of carbon are investigated, since each will be shown to contain not less than 12 parts of that element in every molecule. Furthermore, it is evident that the relative atomic weight of chlorine cannot be a multiple of 35.5, because we have one of the above compounds (methyl chloride) which contains but 35.5 parts of chlorine in each molecule. It follows from this that methylen chloride, methin chloride, and carbon tetrachloride contain, in each molecule, two, three, and four atoms of chlorine respectively. Reasoning from the specific gravities of methane and of the chlorinated methanes, when those substances are converted into gases, we must come to the conclusion that their molecules have the formulæ CH_4, CH_3Cl, CH_2Cl_2, $CHCl_3$, and CCl_4.

In Substituting Hydrogen by Chlorine, the Essential Character of Methane is not changed. In substituting hydrogen by chlorine in methane, the essential character of the latter is preserved. All the chlorinated compounds contain one atom of carbon in each molecule. The hydrogen has been substituted while the carbon has remained. It seems reasonable to suppose, therefore, that in a molecule of methane, it is the carbon atoms which hold the four hydrogen atoms in chemical union with it. In producing the chlorine substitution products, the substituting chlorine atoms have taken the place of the hydrogen atoms which are substituted. We can express this conclusion in the following way: —

The Number of Atoms with which one Atom of Carbon can combine in one Molecule is never more than Four. One atom of carbon is capable of uniting with four atoms of hydrogen, with three atoms of hydrogen and one of chlorine, with two atoms of hydrogen and two of chlorine, with one atom of hydrogen and three of chlorine, and, finally, with four chlorine atoms. In short, one atom of carbon can combine with four atoms of hydrogen, or with any other four atoms which are in combining power equivalent to hydrogen. The number of single atoms with which one atom of carbon can combine in one molecule *is never more than four*.

Valence. Applying what we have learned with carbon to the hydrogen compounds of other elements, we draw the following conclusions: One atom of hydrogen combines with three of nitrogen to form ammonia; two atoms of hydrogen combine with one of oxygen to form water; and one atom of hydrogen combines with one of chlorine to form hydrogen chloride. We see, then, that the individual atoms of the four elements, carbon, nitrogen, oxygen, and chlorine, unite each with a different number of hydrogen atoms, or, to use another term, they have a different *value* toward hydrogen. This we call a difference in *valence;* and we say chlorine has a valence of *one* toward hydrogen, oxygen of *two*, nitrogen of *three*, and carbon of *four*.

This distinction can be expressed in chemical formulæ by writing the symbol for each atom in one molecule separately, grouping those of the hydrogen atoms round the atom of the not-metal, which is supposed to join the different atoms in the molecule. Thus we have —

			H	H	H
H Cl	HOH	HNH	HCH	HCH	HC Cl
		H	H	Cl	Cl
Hydrogen chloride.	Water.	Ammonia.	Methane.	Methyl chloride.	Methylen chloride, etc.

Valence of Carbon, Nitrogen, Oxygen, or Chlorine may not be the same toward all Elements. We must not suppose that each atom of chlorine, oxygen, nitrogen, or carbon can combine with as many atoms of every other element as with those of hydrogen or of chlorine. That this is not the case is shown by the fact that, so far as we know, one atom of carbon can combine with no more than two atoms of oxygen, provided no other elements are present. What has been said in regard to valence, therefore, must be confined exclusively to chlorine and hydrogen compounds.

For further illustration, the formulæ and composition of all the gaseous or gasifiable compounds which we have encountered, are given in the following table: —

	FORMULA.	COMPOSITION BY WEIGHT.
Hydrogen chloride	HCl	One part of hydrogen to 35.5 of chlorine.
Water	H_2O	Two parts of hydrogen to 16 of oxygen.
Ammonia	H_3N	Three parts of hydrogen to 14 of nitrogen.
Methane	H_4C	Four parts of hydrogen to 12 of carbon.
Methyl chloride	H_3ClC	Three parts of hydrogen to 12 of carbon, 35.5 of chlorine.
Methylene chloride	H_2Cl_2C	Two parts of hydrogen to 12 of carbon, 71 of chlorine.
Methin chloride	HCl_3C	One part of hydrogen to 12 of carbon, 106.5 of chlorine.
Carbon tetrachloride	Cl_4C	Twelve parts of carbon, 142.0 of chlorine.
Sulphur dioxide	SO_2	Thirty-two parts of sulphur to 32 of oxygen.
Sulphur trioxide	SO_3	Thirty-two parts of sulphur to 48 of oxygen.
Carbon monoxide	CO	Twelve parts of carbon to 16 of oxygen.
Carbon dioxide	CO_2	Twelve parts of carbon to 32 of oxygen.

From this table it is apparent that in one molecule of carbon monoxide or carbon dioxide we have twelve parts of carbon, just as we have in methane and in the chlorinated methanes. Therefore, by these oxides also, the maximum atomic weight of carbon is fixed at twelve. Furthermore, we see that the least quantity of oxygen occurring in any of the above compounds is 16, and that in all the oxides mentioned we have either 16 parts of oxygen, or some multiple of 16. In the same way the maximum atomic weight of sulphur is fixed at 32, because we have 32 parts of sulphur in one molecule of sulphur dioxide and of sulphur trioxide.

It has been demonstrated that the number of oxygen atoms with which one carbon atom can unite in either carbon monoxide or carbon dioxide is different from the number of hydrogen atoms with which one carbon atom can unite in methane. If we compare carbon dioxide and methane, we see that two hydrogen atoms have exactly the same combining power as one oxygen atom. This difference is not confined to the compounds of carbon alone. With all elements one oxygen atom is equal in combining power to two hydrogen atoms.

The Burning of Methane is a Process of Substituting Hydrogen by Oxygen. The burning of methane, forming water and carbon dioxide, is analogous to the substitution of hydrogen by chlorine. In this case, however, two atoms of oxygen take the place of four atoms of hydrogen, while in substituting with chlorine, one atom of chlorine takes the place of one atom of hydrogen. This comparison is made more evident by the follow- formulæ:—

$$CH_4 \quad + \quad 2\,O_2 \quad = \quad CO_2 \quad + 2\,H_2\,O.$$

1 mol. of methane + 2 mols. of oxygen = 1 mol. of carbon dioxide + 2 mol. of water.

$$CH_4 \quad + \quad 4\,Cl_2 \quad = \quad C\,Cl_4 \quad + \quad 4\,H\,Cl.$$

1 mol. of methane 4 mols. of chlorine = 1 mol. carbon tetrachloride 4 mols. hydrogen chloride.

The action of chlorine on hydrogen compounds is, therefore, analogous to the processes attending the combustion of the same substances in oxygen. In the first case the chlorides may be produced, and in the second the oxides. In either case heat is evolved during the process, and the resulting compounds possess less chemical energy than the ones used for the substitution.*

The Valence of an Element toward Hydrogen is Constant, while it may be Variable toward Oxygen. Finally, the above table shows us that the number of oxygen atoms

* In some cases of the action of chlorine or of oxygen on hydrogen compounds, *substitution* does not take place, but the not-metal is liberated. This is the case with the action of chlorine or of oxygen on ammonia. When chlorine reacts with ammonia, hydrogen chloride and nitrogen are produced. When oxygen reacts with ammonia, water is formed, and nitrogen is liberated. The two processes are therefore analogous.

with which one atom of a given element can unite may vary, while experience has shown us that this is not the case with the hydrogen compounds. *The valence of an element toward hydrogen seems to be constant, while toward oxygen it assumes different values, according to circumstances.*

We have encountered one exception to the above rule. In ammonia one atom of nitrogen unites with three atoms of hydrogen in each molecule; but, as we have seen (page 98), ammonia can add hydrogen chloride to form ammonium chloride. We have demonstrated that in ammonium chloride there are, one atom of nitrogen, four of hydrogen, and one of chlorine in each molecule. It seems possible, therefore, for one nitrogen atom to increase the number of hydrogen atoms with which it is united to four, provided there is some not-metallic element or group of elements (such as chlorine) taken into the combination at the same time. Toward hydrogen alone, nitrogen seems to have an invariable valence of three, in compounds containing but one atom of nitrogen in each molecule. Chemically, it is said that ammonia is unsaturated because it can add to itself certain acid compounds in order to produce ammonium salts. In ammonium chloride we have one atom of nitrogen joined to four atoms of hydrogen and one of chlorine; the valence of nitrogen in this compound is consequently five.

Summary.

1. When chlorine acts on most hydrogen compounds of the not-metals, hydrogen chloride is produced, and the not-metal contained in the particular compound in question is liberated.

2. If an excess of chlorine acts violently on methane (i.e., when a mixture of methane and chlorine is ignited), carbon is liberated, and hydrogen chloride is produced.

3. If chlorine acts slowly upon methane, substitution takes place in such a way that one atom of chlorine always takes the place of one atom of hydrogen. At the same time one molecule of hydrogen chloride is formed.

4. The four hydrogen atoms in one molecule of methane can be successively substituted by chlorine, producing substances having the chemical formulæ CH_3Cl, CH_2Cl_2, $CHCl_3$, and CCl_4.

5. The specific gravities of methane and of its substitution products, when the latter are heated to a temperature high enough to convert them into gases, confirm the formulæ given under 4.

6. The atomic weights of carbon and of chlorine, determined by the specific gravities of gaseous methane and its substitution products, are 12 and 35.5 respectively.

7. The number of single atoms with which one atom of carbon can combine in any one molecule is never greater than 4.

8. One atom of chlorine, oxygen, nitrogen, or carbon can unite with different numbers of atoms of hydrogen. This is termed a difference in valence. Chlorine has a valence of one toward hydrogen, oxygen of two, nitrogen of three, and carbon of four.

9. The number of oxygen atoms with which one atom of these elements can unite is different from the number of hydrogen atoms. This difference is such that one oxygen atom can take the place of two hydro-

gen atoms, and is shown in the table comparing the formulæ of the gaseous compounds discussed in this work.

10. The number of oxygen atoms with which one atom of the above elements can unite may vary. This is shown by the existence of two oxides of sulphur, sulphur dioxide and sulphur trioxide, and of the two oxides of carbon, cabron monoxide and carbon dioxide.

CHAPTER XXIII.

THE FORMATION OF SALTS BY DOUBLE DECOMPOSITION.

Formation of Salts by Substitution and Neutralization. Thus far in considering the formation of salts we have found two methods available for the purpose, — one by the action of metals on acids (see page 45 and 57), the other by neutralization of acids by bases (see pages 48 and 58). In both cases the salt is formed by a process of substitution in which the metal, either alone or in combination with oxygen, takes the place of the hydrogen in the acid. The salt is therefore the acid in which the whole, or a part, of the hydrogen has been replaced by some other metal. The acids, therefore, may be considered as *salts of the metal hydrogen*. Thus, sulphuric acid is hydrogen sulphate, hydrochloric acid is hydrogen chloride, and nitric acid is hydrogen nitrate; while the salts of these acids are termed sulphates, chlorides, and nitrates. The parallelism can be made clearer by comparing the following formulæ:—

H_2SO_4.	Na_2SO_4.
Hydrogen sulphate (sulphuric acid).	Sodium sulphate.
HCl.	$NaCl$.
Hydrogen chloride (hydrochloric acid).	Sodium chloride.
HNO_3.	$NaNO_3$.
Hydrogen nitrate (nitric acid).	Sodium nitrate.

Salt Formation by Action of an Acid on a Salt. There are, however, other means of salt formation of as great importance chemically as those just mentioned which do not involve the action of an acid on a base.

If sodium chloride is brought in contact with an excess of concentrated sulphuric acid, a gas passes off, which from its properties can easily be recognized as hydrogen chloride (see page 38 and Experiment 16, Appendix). After the reaction is completed, if the remainder is dissolved in water and the solution evaporated, crystals of a salt, which differs from sodium chloride, will separate. This salt does not liberate hydrogen chloride when sulphuric acid is added to it. It melts when heated, and finally gives off dense fumes of sulphuric acid. It then again solidifies, to melt once more at a bright white heat. In short, it has all of the properties belonging to primary sodium sulphate (see Experiment 34, Appendix). From sodium chloride and hydrogen sulphate we have therefore produced sodium sulphate and hydrogen chloride. This change can be made plain by the following chemical formulæ: —

$$NaCl + HHSO_4 = HCl + NaHSO_4.$$

A reaction of this kind is termed a *double decomposition*, since both the sulphuric acid and the sodium chloride have been altered in such a way that sodium and hydrogen have exchanged places. Hydrogen chloride is a gas, and, as a consequence, passes off, so that at the end of the experiment nothing remains but primary sodium sulphate, together with the excess of sulphuric acid which has been used. In such a reaction the action of sodium chloride is analogous to that of sodium hydroxide on sulphuric acid; in one case sodium chloride and sulphuric acid produce sodium sulphate and hydrogen chloride; in the other, sodium hydroxide and sulphuric acid form sodium sulphate and hydrogen oxide (water). It follows from this that the neutralization of an acid by a base is also a reaction involving a *double decomposition.*

Double Decompositions between Two Salts. The reactions which have been cited involve the action of a salt or a base on an acid. We can, however, have other double decompositions, in which two salts change each other so as to produce two new salts, in which case the metals exchange places. If, for example, a solution of barium chloride is brought in contact with one of sodium sulphate, a double decomposition takes place, while the insoluble barium sulphate is separated. This change can be expressed as follows:—

Barium chloride +	Sodium sulphate =	Barium sulphate +	Sodium chloride.
soluble	soluble	insoluble	soluble.

The separation of an insoluble substance from a solution by double decomposition is termed *precipitation.* Processes similar to the above are of such frequent occurrence in chemical work as to warrant the establishment of the following arbitrary rule:—[75]

Double decompositions occur whenever a volatile or an insoluble substance can be produced.

EXAMPLES:—

1. Calcium chloride +	Potassium carbonate =	Calcium carbonate +	Potassium chloride.
soluble	soluble	insoluble	soluble.
2. Barium chloride +	Sodium carbonate =	Barium carbonate +	Sodium chloride.
soluble	soluble	insoluble	soluble.
3. Sodium chloride +	Silver nitrate =	Silver chloride +	Sodium nitrate.
soluble	soluble	insoluble	soluble.
4. Magnesium sulphate +	Barium chloride =	Barium sulphate +	Magnesium chloride.
soluble	soluble	insoluble	soluble.

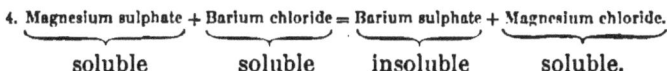

In the above changes, the result is such that two soluble salts so react that they completely decompose each other in order to produce two new salts, one of which is insoluble in the medium in which the reaction takes place. Of course, if any substance is added to the solvent which will prevent the formation of an insoluble substance, then precipitation cannot take place. For example, the carbonates are decomposed by acids, with the liberation of carbon dioxide and water. It follows that reactions 1 and 2 cannot take place if an acid be present, as the carbonate of calcium which should be formed would be decomposed by this acid. If, however, in reaction 2 the acid were sulphuric acid, then a precipitation of the insoluble barium sulphate would take place, instead of that of the equally insoluble barium carbonate.[76]

Causes of Double Decomposition. Electrolysis. The reason why such double decompositions take place is probably to be found in the following explanation: —

If the two poles of a battery are placed in a salt solution, the salt will be decomposed by the electric current which is conducted from one pole to the other. Of these two poles, one is positive and the other negative. The current passes from the negative to the positive pole. Substances which, like salts, are decomposed by an electric current are termed *electrolytes.* The terminals from which the current enters or leaves the solution are called the *electrodes.* The positive electrode is the *anode,* and the negative one the *kathode.* The portions into which the electrolyte separates are termed the *ions.* The ion which passes off at the positive pole (itself negative) is the *anion,* that passing off at the negative pole (itself positive) is termed the *kation.* For example, hydrochloric acid is separated by means of an electric current into its ions, hydrogen and chlorine. The kation is hydrogen, the anion is chlorine; and, if proper means are taken to collect the gases, pure hydrogen can be obtained from the negative pole and chlorine from the positive, as was done in the experiments with hydrochloric acid in Experiment 10, Appendix. In the same way a solution of sodium chloride is separated into sodium (the kation) and chlorine (the anion). A solution of potassium chloride will separate into potassium (the kation) and chlorine (the anion), or a solution of potassium sulphate (chemical formula K_2SO_4) into potassium (kation) and the group SO_4 (anion).* Only such solutions as contain substances capable of separation into their ions have the power of conducting electricity. Hence it seems probable that salts separate, at least partially, when they are dissolved, even if no electric current is passed through them. This supposition is borne out by a number of other experimental facts of too complex an order to be mentioned here.

A solution of sodium chloride contains the ions represented by the symbols Na and Cl, one of potassium chloride, K and Cl, one of potassium sulphate, 2 K and SO_4, and so on. We cannot, however, demonstrate the presence of the uncombined ions in such a solution (i.e., we cannot prove the presence of free chlorine and free sodium in a solution of sodium chloride) so long as these elements are not collected at electrodes by the action of an electric current.

It can be supposed, for example, that, when sodium chloride separates into Na and Cl, the individual atoms do not remain permanently isolated, but that they frequently recombine. Each atom of chlorine, however, does not unite with the same atom of sodium with which it was originally joined, but with another, and again separates and rejoins others in the throng of atoms which it may encounter.

Change in the Conditions if an Insoluble Substance can be formed. According to this theory, if a solution of sodium chloride is mixed with one of potassium sulphate, we shall have present the ions of Na + Cl and 2 K + SO_4. Such a mixture is obviously practically identical with one of potassium chloride and sodium sulphate, for the

* Of course, the potassium or sodium, being in the presence of water, will react again with that liquid. The metals themselves can only be separated by electrolyzing the fused salts, when the water is present.

latter contains ions K + Cl and 2 Na + SO_4. The result would then be the same as if on mixing potassium chloride with sodium sulphate we were to have partial double decomposition forming sodium chloride and potassium sulphate, both of which are soluble in water. An entirely different effect is obtained if one of the salts produced is *insoluble.* For example, silver nitrate in solution with sodium chloride undergoes double decomposition, forming the insoluble silver chloride and the soluble sodium nitrate. Now, in a solution of silver nitrate we have the ions Ag * and NO_3 (the chemical formula of silver nitrate is Ag NO_3), and in a solution of sodium chloride we have the ions Na and Cl. The chemical structure of such a solution could be represented by the formula —

$$\underset{\text{kation}}{Na} + \underset{\text{anion}}{Cl} + \underset{\text{kation}}{Ag} + \underset{\text{anion.}}{NO_3}.$$

The kation Ag is capable of uniting with the anion Cl, when it comes in contact with it, and the resulting compound (Ag Cl, silver chloride) is insoluble. It is, therefore, precipitated, and separates from the solution. A second kation Ag can come in contact with a second anion Cl, and so on until all of the silver and chlorine present are separated as silver chloride. The solution will then contain only the ions Na and NO_3†, which, after the solvent is evaporated, will leave sodium nitrate. What is true of the above reaction must be true in every case where salts in solution are brought

* The chemical symbol for one atom of silver is Ag, from the latin *argentum* (silver).

† Of course if more sodium chloride than is necessary to form silver chloride with all of the silver in solution is present, then that excess remains after the double decomposition, and *vice versa.*

in contact, when *one of the products of the union of the ions is insoluble.* If one of the constituents is volatile, it is just as surely removed from the solution as if it were insoluble. From the above considerations it follows that complete double decomposition must take place whenever insoluble or a volatile substance can be produced.

Electrolysis of Water and of Hydrochloric Acid. The process of electrolysis has been repeatedly used during the progress of this work. Thus water acidulated with sulphuric acid was decomposed into hydrogen and oxygen by the electric current (see page 21). Pure water is not a conductor of electricity, and consequently is not subject to electrolysis. But when sulphuric acid is dissolved in water the mixture becomes a conductor, and if an electric current is passed through it, the kation H_2 separates at the negative pole, and the anion SO_4 at the positive. But SO_4 is not capable of existence as such; it breaks down into sulphur trioxide (SO_3, see page 56) and oxygen, which passes off. The sulphur trioxide, being the anhydride of sulphuric acid, dissolves in water to form sulphuric acid, so that no loss of the latter substance takes place. The only compound which is decomposed is therefore *water.*

This change may be represented as follows by chemical symbols:—

$$H_2SO_4 = H_2 + SO_4.$$
Sulphuric acid = kation + anion.

$$SO_4 = SO_3 + O.$$
The anion = sulphur trioxide + oxygen.

$$SO_3 + H_2O = H_2SO_4.$$
Sulphur trioxide + water = sulphuric acid.

The substances passing off are hydrogen and oxygen.

A concentrated solution of hydrochloric acid is decomposed into hydrogen and chlorine by the electric current (see page 40). In this case the kations of hydrogen separate at the negative pole and the anions of chlorine at the positive.

Neutralization of Bases by Acids. To illustrate the neutralization of bases by acids, let us take the neutralization of hydrochloric acid by sodium hydroxide. This change can be represented by the chemical formulæ —

$$\underset{\text{Hydrochloric acid}}{HCl} + \underset{\text{Sodium hydroxide}}{NaOH} = \underset{\text{Sodium chloride}}{NaCl} + \underset{\text{Water}}{HOH}.$$

In solution, hydrochloric acid exists as the ions $H + Cl$, and sodium hydroxide as the ions $Na + OH$. When these are brought in contact the ions would be:—

$$H + Cl + Na + OH.$$

Now, the only substance which is present in excess, and which, while being capable of formation from the above constituents, would not have a tendency toward ionization, is *water*. Water, therefore, would be produced by the interaction of these ions, while the ions $Na + Cl$ alone would be left. The same result is obtained when other acids and bases are brought in contact; so that one of the chief reasons for the fact that acids and bases neutralize each other is found in the formation of water. Of course the neutralization would be the more rapid and complete the more readily ionization took place on solution.

The process of electrolysis has been a most important one in the history of chemistry. By means of an electric current, Davy first succeeded in decomposing caustic potash (potassium hydroxide) and caustic soda (sodium hydroxide), and in isolating the metals potassium and sodium. Before that time caustic potash and caustic soda were considered elements.

Summary.

1. Acids are salts of the metal hydrogen, and salts are acids in which the hydrogen has been replaced by other metals.

2. Salts of some acids can be formed by the action of those acids upon the salt of some other acid; for example, by the action of sulphuric acid on sodium chloride we can produce sodium sulphate.

3. Such reactions are termed double decompositions, for in them the metal of the salt and the hydrogen of the acid acting upon it exchange places.

4. The neutralization of a base by an acid is also a double composition.

5. Double decomposition can also take place between two salts. In such a case the metals in the two salts exchange places. Complete double decompositions occur if a volatile or an insoluble substance is formed during the reaction.

6. Salt solutions are decomposed (electrolyzed) by the electric current. The salt separates into two ions, — the kation (metallic constituent) and the anion (not-metallic constituent). Only such solutions as contain substances separable by electrolysis conduct electricity.

7. Salts, when dissolved, separate into their ions, at least in part. The presence of these ions cannot be demonstrated by ordinary means. They can, however, be collected at the electrodes by the action of the electric current.

8. When, on mixing two salts, a new insoluble salt can be produced, then a precipitate of such salt is formed. The remaining two ions stay in the solution and form the second new salt on evaporation of the solvent. The case is the same if one of the substances produced is volatile.

9. In electrolyzing water acidulated with sulphuric acid, the acid is decomposed into its ions, hydrogen and a group of elements represented by the formula SO_4. The latter separates at the positive pole, and breaks down into sulphur trioxide (SO_3) and oxygen. The elements which pass off are, therefore, hydrogen and oxygen. The sulphur trioxide remains, and, with the water which is present, regenerates sulphuric acid.

10. The process of neutralization is explained by the decomposition of the acid and the base into their ions. Two of these ions must always be hydrogen and hydroxyl (H and OH), and these unite to form water. There would then remain in solution only the metallic and not-metallic ions which were originally present as constituents of the base and acid respectively. These ions, when the solvent is evaporated, unite to form the salt.

CHAPTER XXIV.

THE CHEMICAL NATURE OF SOME OTHER ELEMENTS AND COMPOUNDS RELATED TO THOSE WHICH HAVE BEEN STUDIED.

Chemical Elements are Separable into Groups and Families. The chemical elements are not individuals each with distinct characteristics which do not recur in the natural history of the others. They are members of certain groups or families, the representatives of which are all more or less related to each other, both chemically and physically. To trace out these relations in their entirety would involve too great extension of the limits of an elementary work. We can call attention only to a few individuals which are closely related to the elements which we have studied.

Elements of the Chlorine Family. Chlorine is very closely connected, chemically, with three other elements, —fluorine, bromine, and iodine. These substances, like chlorine, are always found in nature in combination with metals, as fluorides, bromides, and iodides. The bromides and iodides of potassium and sodium are generally found in conjunction with the chlorides of the same metals, but they are present in lesser quantity on the earth's surface. In sea-water there are large amounts of sodium chloride and small amounts of sodium bromide and iodide. All the elements of the chlorine family, with the exception of fluorine, can be isolated from their salts by the same means.

Of these four elements, fluorine has the smallest atomic weight, chlorine the next higher, bromine the next, and iodine the greatest. In the following table these elements are given in the order of their increasing atomic weights, while the physical properties, which vary regularly in the order given, are placed opposite each individual:—

	Atomic Weight.	Specific Gravities as Gases.	Usual Condition.	Color of Vapor.	Boiling-Points.	Melting-Points.	Specific Gravities of Solids.
Fluorine	19.	1.364	Slightly yellow gas	Yellow	—	—	—
Chlorine	35.5	2.46	Yellowish-green gas	Yellow	−35°	−102°	
Bromine	80.	5.54	Dark-brown liquid	Brown	63°	− 7°.3	
Iodine	127.6	8.84	Crystalline black solid	Violet	200°	114°	4.94

As is seen from the above, the specific gravities, both of the gaseous and solid substances, increase with increasing atomic weight, the color deepens, and finally (with iodine) becomes black. The boiling and melting points also increase with increasing atomic weight.

Elements of the Oxygen Family. Oxygen and sulphur also belong to a family which contains four elements. In order to show the similarity to the chlorine family in the changes brought about by the increase in atomic weight, a table containing the principal physical properties of these elements is given on opposite page.

Elements of the Nitrogen Family. In the family of which nitrogen is a member, we have five individuals,

	Atomic Weight.	Specific Gravity as Gases.	Usual Condition.	Color of Vapor.	Boiling-Points.	Melting-Points.	Specific Gravities of Solids.
Oxygen	16.	1.105	Colorless gas	Colorless	− 182°	—	—
Sulphur	32.	2.2	Yellow solid	Brown	440°	114°	2.045
Selenium	78.	5.7	Dark-brown, almost black solid	Brown	665°	217°	4.5
Tellurium	125.	9.	Metallic appearing solid	Orange		500°	6.25

—nitrogen, phosphorus, arsenic, antimony, and bismuth. Of these, the element with the smallest atomic weight (nitrogen) is completely a not-metal, while that with the highest (bismuth) is a metal, physically and chemically. In this family, therefore, we have the same change in properties, with increasing atomic weight, as was observed in the families of which chlorine and oxygen are representatives. This fact will be plainer by consulting the table:—

	Atomic Weight.	Specific Gravity as Gases.	Usual Condition.	Color of Vapor.	Boiling-Points.	Melting-Points.	Specific Gravity of Solids.
Nitrogen	14	.972	Colorless gas	Colorless	− 193	− 203°	—
Phosphorus	31	4.16	Yellow solid	Colorless	290°	44°	1.83
Arsenic	75	10.3	Steel-gray solid	Lemon yellow	450° *	—	5.76
Antimony	120	9.78 †	Silver white solid	—	About 1200°	425°	6.7
Bismuth	208.9	10.1	Reddish metallic solid	—	1400° (?)	270°	9.93

* Arsenic vaporizes without previously melting.

† The specific gravity of antimony vapor is less than that of arsenic. This is because arsenic, when a gas, has molecules formed of four atoms, while antimony, as a gas, has molecules formed of two or less. The molecules of gaseous bismuth consist, in part, of but one atom, hence its low specific gravity.

In all these families which have been mentioned, the increase in atomic weight, as we pass from member to member, brings with it an increase in specific gravity both as a vapor and as a solid. There is also a gradual change from a not-metallic to a metallic element, or, at least in the chlorine family, to an element which is more like a metal than the original member.

Comparison of the Three Families. As a rule then, in these three families, the specific gravities of the elements, both when taken as gases and as solids, increase as the atomic weight increases, and at the same time the melting-points become higher. The elements with the lowest atomic weights are not metallic, while those with the highest are either metallic in appearance (tellurium and arsenic, antimony and bismuth), or at least approach the metallic character (iodine). In chemical character we find the same close family relationship.

Chemical Resemblances in the Hydrogen Compounds. Our attention, in the preceding portions of this work, has mainly been directed to a consideration of compounds of the elements with *hydrogen*. Therefore, in calling attention to the chemical resemblances of the members of the three families under discussion, we shall use the hydrogen compounds as a basis.

Hydrogen Compounds of the Chlorine Family. Fluorine, chlorine, bromine, and iodine are capable of producing one hydrogen compound each; and this, like hydrogen chloride, is a gas at ordinary temperatures.* These hydrogen compounds are all readily soluble in water,

* Hydrogen fluoride boils at 19°, and is, therefore, a gas at summer heat, but a liquid at winter temperature.

and their solutions are acids, capable of neutralizing bases to form salts. These substances are termed hydrofluoric, hydrochloric, hydrobromic, and hydroiodic acids. The salts derived from the acids by replacing the hydrogen by other metals are the fluorides, chlorides, bromides, and iodides. These hydrogen compounds are all formed on the same structural plan; i.e., by the union of equal volumes of hydrogen and of the not-metallic element (fluorine, chlorine, bromine, or iodine), so that their smallest particles contain one atom of hydrogen and one atom of fluorine,* chlorine, bromine, or iodine. The chemical formulæ of these substances are, therefore —

HF, the chemical symbol for one atom of fluorine being F.
H Cl, the chemical symbol for one atom of chlorine being Cl.
H Br, the chemical symbol for one atom of bromine being Br.
HI, the chemical symbol for one atom of iodine being I.

The stability of these hydrogen compounds diminishes as the atomic weight of the not-metallic element increases. As a consequence, hydrogen iodide is much more readily decomposed than hydrogen bromide, and hydrogen bromide more readily than hydrogen chloride. Hydrogen fluoride has not been separated into its elements by heat alone. Hydrogen chloride is decomposed only at a high white heat, while hydrogen iodide can be broken down by a red-hot wire placed in the gas. The relative stability of the last three of these compounds is readily shown by the following experiments: —

If a few drops of a solution of chlorine in water are added to a solution of hydriodic acid, iodine will at once separate and hydro-

* The specific gravity of hydrogen fluoride, as a gas, shows that this substance has a molecule composed of two atoms of hydrogen and two of fluorine. Its formula, therefore, being $H_2 F_2$.

chloric acid be formed. Chlorine, therefore, expells iodine from its hydrogen compound, to form the more stable hydrochloric acid. In the same way bromine is replaced by chlorine, when the latter element is brought in contact with hydrobromic acid; and lastly, iodine is expelled from hydroiodic acid by bromine.[77]

Hydrogen Compounds of Elements of the Oxygen Family. The elements of the family of which oxygen is a member all form hydrogen compounds which, with the exception of hydrogen oxide (water), are colorless gases at ordinary temperatures. These hydrogen compounds are all of the same structural plan, for their molecules are formed of two atoms of hydrogen and one atom of the not-metallic element. This will be seen from the following chemical formulæ: —

H_2O, water.
H_2S, hydrogen sulphide.
H_2Se, hydrogen selenide (the chemical symbol for one atom of selenium is Se).
H_2Te, hydrogen telluride (the chemical symbol for one atom of tellurium is Te).

In regard to their relative stability, these hydrogen compounds follow the same rule as in the case of chlorine, bromine, and iodine; the higher the atomic weight of the element forming them is, the more easily are they decomposed. Thus, water is broken down into hydrogen and oxygen only at a high white heat, while hydrogen telluride is so unstable that it decomposes in part, even at ordinary temperatures. The compounds hydrogen sulphide and hydrogen selenide lie between these two extremes.

Hydrogen Sulphide, Preparation and Properties. Next to water, the most important hydrogen compound of the elements of the oxygen family is hydrogen sulphide. This substance is so frequently used as a laboratory reagent that a brief description of its properties and its reactions is necessary.

The preparation of hydrogen sulphide involves the same principle as that of hydrogen chloride. The sulphide of a metal is treated with an acid, just as sodium chloride is brought in contact with sulphuric acid, in order to liberate hydrogen chloride; for example: —

Sulphide of iron (ferrous sulphide) with hydrochloric acid forms chloride of iron (ferrous chloride) and hydrogen sulphide.

Sulphide of zinc with sulphuric acid forms zinc sulphate and hydrogen sulphide.

All metallic sulphides, however, are not decomposed by acids, there being a considerable number which are not acted upon. In the laboratory preparation of hydrogen sulphide, it is advisable to decompose ferrous sulphide with dilute sulphuric acid, producing ferrous sulphate and hydrogen sulphide.[78] In chemical notation this reaction can be expressed as follows: —

$$\underset{\text{Ferrous sulphide}}{FeS} + \underset{\text{Sulphuric acid}}{H_2SO_4} = \underset{\text{Ferrous sulphate}}{FeSO_4} + \underset{\text{Hydrogen sulphide.}}{H_2S.}$$

This reaction, therefore, involves a double decomposition, and is entirely analogous to the neutralization of a base by an acid, as the following reaction will demonstrate: —

$$\underset{\text{Ferrous oxide (a base)}}{FeO} + \underset{\text{Sulphuric acid}}{H_2SO_4} = \underset{\text{Ferrous sulphate}}{FeSO_4} + \underset{\text{Water.}}{H_2O.}$$

Properties of Hydrogen Sulphide. Under ordinary conditions, hydrogen sulphide is a gas with a most disagreeable odor. When inhaled, even in comparatively small quantities, it is poisonous. It is soluble in water in considerable quantity (one cubic centimetre of water dissolves three of hydrogen sulphide at 18°); and its solution, when allowed to stand exposed to the air, gradually deposits sulphur, owing to the fact that it is oxidized. This change can chemically be repeated as follows:—

$$\underset{\text{Hydrogen sulphide}}{H_2S} + \underset{\text{Oxygen}}{O} = \underset{\text{Water}}{H_2O} + \underset{\text{Sulphur.}}{S}$$

It naturally follows from the above that hydrogen sulphide burns in the air just as does methane. When methane burns, it forms carbon dioxide and water; when hydrogen sulphide burns in an excess of oxygen, it forms sulphur dioxide and water.

Formation of Insoluble Sulphides by Double Decomposition between Hydrogen Sulphide and the Salts of Metals. Most sulphides of the metals are not soluble in water; and, as has been mentioned above, a considerable number are not decomposed by dilute acids. Consequently, in many cases, the insoluble sulphides are precipitated from solutions of metallic salts by the addition of hydrogen sulphide.[79] For example:—

$$\underbrace{\text{Copper sulphate}}_{\text{Soluble}} + \underbrace{\text{Hydrogen Sulphide}}_{\text{soluble}} = \underbrace{\text{Copper sulphide}}_{\text{insoluble}} + \underbrace{\text{Sulphuric acid.}}_{\text{soluble.}}$$

$$\underbrace{\text{Lead acetate}}_{\text{Soluble}} + \underbrace{\text{Hydrogen sulphide}}_{\text{soluble}} = \underbrace{\text{Lead sulphide}}_{\text{insoluble}} + \underbrace{\text{Acetic acid.}}_{\text{soluble.}}$$

Some Sulphides Insoluble in Water are Soluble in Dilute Acids. The sulphides of certain metals, although they are insoluble in water, are decomposed by dilute acids, and hence are soluble in the latter. This is the case with ferrous sulphide or zinc sulphide which were mentioned above. Under these circumstances, it is obvious that no precipitate, or at least very little precipitate, of the sulphide is produced by hydrogen sulphide; for, as the action of hydrogen sulphide in producing the sulphide of the metal liberates an acid, it follows that the latter would reverse the reaction as soon as it is formed. In other words, it would attack the sulphide, and reproduce the original salt of the metal together with hydrogen sulphide. For example:—

Zinc sulphate +	Hydrogen sulphide =	Zinc sulphide +	Sulphuric acid.
Soluble	soluble	insoluble	soluble.

After a certain quantity of sulphuric acid has been formed, the latter attacks the zinc sulphide in order to reproduce zinc sulphate and hydrogen sulphide. If care is taken to neutralize the acid as fast as it is formed, the precipitation is complete. The neutralization can be accomplished by adding the sulphide of an alkali metal instead of hydrogen sulphide, thus:—

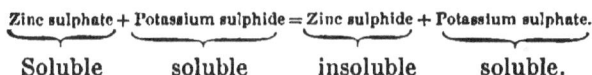

Zinc sulphate +	Potassium sulphide =	Zinc sulphide +	Potassium sulphate.
Soluble	soluble	insoluble	soluble.

In this case no free sulphuric acid is formed, and hence no solution of the zinc sulphide takes place.[80] The chemical importance of hydrogen sulphide is mainly found in its use as a means of separating certain groups of metals from each other when they are found in solution as salts. If hydrogen sulphide is added to the acid solution of the salts of a number of metals, some of which form sulphides insoluble in dilute acids, while others produce soluble sulphides, it is obvious that the metals of the first class can be completely separated by the addition of the hydrogen sulphide, while those of the second remain dissolved. After the hydrogen sulphide has separated all that it will, the precipitate can be filtered off, and then the soluble sulphides can be separated by the neutralization of the acid by a base. By this means a complete separation of the two classes of metals can be effected, and such separations are the basis of the branch of chemistry termed *analytical chemistry.*[81]

The Hydrogen Compounds of the Elements of the Nitrogen Family. The elements of the nitrogen family, with the exception of bismuth, are capable of forming hydrogen compounds which are constructed on the same plan as ammonia; i.e., for one atom of the characterizing element in each molecule of the hydrogen compound, we have three atoms of hydrogen:—

Ammonia (hydrogen nitride), chemical formula, NH_3.
Phosphine (hydrogen phosphide), chemical formula, PH_3.*
Arsine (hydrogen arsenide), chemical formula, $As\,H_3$.
Stibine (hydrogen antomonide), chemical formula, $Sb\,H_3$.

Like the hydrogen compounds in the other families, those in this diminish in stability as the atomic weight of the element entering into their formation increases. Ammonia is quite stable, while arsine and stibine are both very easily decomposed into their constituent elements by heating.

Lastly, in the carbon family, we also have five elements, — carbon, silicon, germanium, tin, and lead. We have here the same changes with increasing atomic weight. Carbon is a not-metal; lead, with the highest atomic weight, is a metal. Only carbon and silicon are capable of forming compounds with hydrogen, and the hydrogen compound of silicon, while much less stable than methane, is formed according to the same plan:—

Methane, chemical formula, CH_4.
Silicon hydride, chemical formula, $Si\,H_4$.

* The chemical symbols for one atom of phosphorus, arsenic, and antimony are P, As, and Sb respectively. The symbol Sb for antimony is derived from the Latin *stibium.*

The few relationships which have been mentioned will serve to show what is meant by the statement that "the elements are divided into certain groups or families." What is true of the few elements studied is also true of every other element with which we are acquainted. Each one of these is a member of a family with several representatives which are closely related chemically and physically. In larger works on chemistry, where the characteristics of all of the elements are considered, the grouping by families forms the basis for the systematic arrangement of such works.

APPENDIX.

PRECAUTIONS TO BE TAKEN WHILE WORKING IN THE LABORATORY.

Before entering upon laboratory work the pupil should read and remember the following cautions: —

Burns, stains, and fire. — Yellow phosphorus should never be handled except with a pair of tongs or pincers. When exposed to the air in a warm room it may take fire spontaneously. If touched by the hand it will take fire. The burns so produced are extremely painful, and may become dangerous by reason of phosphorus poisoning. *Sodium and potassium* are kept under naphtha. Pieces larger than beans should *never* be placed in water; and in any event, a very small piece should first be tested. Sodium which has not a clear and bright surface when cut, should be rejected. In all cases the outer coating of oxide should be cut away before placing the metal in water. Burns are best treated by covering the spot with a solution of cocaine in sweet oil, and then with an emulsion of lime-water, glycerine, and sweet oil. *Nitric acid* stains the skin yellow. Where concentrated, it will cause an ulcer to form. *Bromine* stains the skin brown, and unless instantly removed it will cause a painful ulcer. *Iodine* stains the skin dark violet; *nitrate of silver*, black. Of course every precaution should be taken to keep the above substances from touching the hands or face; but, in case of accident, washing with clean water will be best in the case of nitric acid and bromine. A solution of sodium hyposulphite followed by water will remove iodine. *Concentrated sulphuric acid* will attack the skin; not so rapidly, however, as nitric acid. In case of an accident this can often be removed by washing with water or sodium carbonate solution before serious results have followed. *Hot sulphuric acid* will *instantly* produce the most painful burns. *Any test-tube* in which sulphuric acid or anything else is being heated, should be held by a test-tube holder *with its mouth pointing away from the manipulator or from any one standing near. Ether or carbon bisulphide* must not be used within six feet of a burning Bunsen burner. These liquids take fire with the greatest readiness. Matches should be kept in a tin box, which is never to be placed in the drawer of the desk, but should always be kept outside.

Inhalation of fumes and gases. — Chemical experiments which will develop poisonous or irritating gases should always be performed under a hood with a good draught.* *Chlorine and bromine* attack the mucous membrane of the eyes, throat, and nose. Continued inhalation will give rise to bronchial inflammation. Chlorine or bromine will also cause nausea. If, by accident, the pupil should take an excessive quantity of chlorine into the lungs, the quickest remedy is probably the inhalation of the fumes of alcohol. *The gaseous oxides of nitrogen* are poisonous, causing violent headache and nausea. Phosphine, arsine, and stibine are very poisonous. Ammonia is quite irritating. Work in which these substances are generated or used *must* be done under the hood. *Sulphuric acid* should not be heated to above 150° unless the apparatus is under the hood. The acid will break down, partly into water and sulphur trioxide; the vapors of the latter are irritating to the lungs. Liquids containing *hydrochloric acid* or *nitric acid* should be evaporated under the hood. *Sulphuretted hydrogen* is poisonous and disagreeable; continued inhalation of even small quantities will cause headache, and may have serious results. It is, therefore, imperatively necessary, unless a room is especially provided in which to generate this gas, that all work with hydrogen sulphide should be performed under the hood. Sulphur dioxide is extremely irritating; work with this gas must always be done under the hood.

Explosions. — Most accidents result from carelessness; therefore, the invariable rule by which the student should govern himself in the laboratory is, "Never be careless; for carelessness may

* So urgent is this rule that pupils should be forbidden even to heat test-tubes or small evaporating dishes with reagents which will give off fumes of hydrochloric acid, nitric acid, hydrogen sulphide, bromine, chlorine, nitric oxide, etc., unless they do so under the hood. A good hood is as necessary as a good burner.

result in permanent disfigurement or loss of sight." Hydrogen and oxygen, hydrogen and air, hydrogen and chlorine, gaseous hydrocarbons and oxygen, phosphine and oxygen, or phosphine and air, as well as the other not very stable hydrogen compounds of the not-metals, mixed with oxygen or air, will cause violent explosions when ignited. In generating these gases, extreme care must be taken not to bring a flame near the exit-tube of the apparatus until the pupil is *sure* that a brisk current of the generated gas has traversed the apparatus for sufficient length of time to expel all air. No definite time rule can be established, because this will vary with the size of the apparatus; but, when using the ordinary generating flasks of from 300 to 500 cc., the pupil should wait at least seven to ten minutes. *Chlorate of potassium, permanganate of potassium,* and similar powerful oxidizers must *not* be rubbed in a mortar when in contact with substances which are readily oxidized (sugar, starch, sulphide of antimony, sulphur, phosphorus, etc.). Care must be taken to have all drying trains or other apparatus for washing gases in such a condition that a current of gas can pass through freely; otherwise, when a gas is being generated, the pressure caused by a stoppage may give rise to a dangerous explosion. Safety-tubes must always be open; otherwise they are no longer *safety*-tubes, for their being stopped may cause an explosion. The openings by which gases escape from the generating apparatus must be sufficiently large to allow a free current of gas to flow.

In a well-conducted laboratory, desks and apparatus must always be kept as clean as possible, and *reagent bottles returned* to their proper places as soon as the occasion requiring their use is over. Bunsen burners can be cleaned by unscrewing the outer tube, and brushing the nipple with a dry, stiff, test-tube brush.

LABORATORY EXPERIMENTS.

Carefully read the laboratory precautions on preceding pages.

The following experiments are to be conducted in the laboratory in connection with the work on the text. The pupil must carefully follow all the directions, and answer all questions, putting his observations and answers in a note-book. Some of the experiments can be performed only by the teacher. These are marked "Teacher's experiments." Their description is to be given in the pupil's note-book exactly as if they had been performed by the pupil himself.

---

1. The Thermometer

The thermometer in general scientific use is the centigrade; the one in popular use in the United States is the Fahrenheit. The usual thermometer measures the temperature by means of the expansion of mercury when the temperature is increased, or the contraction when it is diminished. It consists of a capillary tube, one end of which terminates in a bulb. This bulb and a portion of the capillary tube are filled with mercury. The upper part of the tube is either a vacuum, or is filled with nitrogen, and its tip closed by fusing the glass. The bulb is first placed in melting ice, and the position of the mercury column is marked as 0° (in the centigrade thermometer). The thermometer is then transferred to water which is boiling under a pressure of one atmosphere, and the point at which the upper end of the mercury column stands is marked 100°. The interval is divided into 100 equal parts, called degrees. In the Fahrenheit thermometer the melting-point of ice is marked 32°, and the boiling-point of water, 212°. The divisions in each thermometer are carried some distance below 0°, and these are designated by the minus sign. The diagram, Fig. 1, shows the difference between the centigrade and Fahrenheit thermometers.

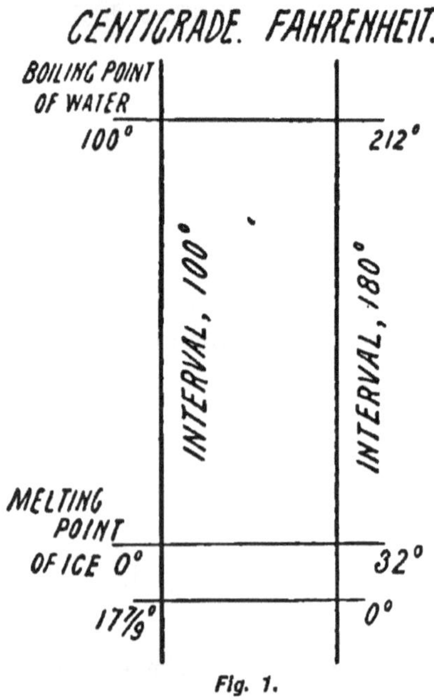

Fig. 1.

The fundamental difference between the thermometers is that

100° Centigrade = 180° Fahrenheit,

or 5° " = 9° "

The temperature centigrade in terms of Fahrenheit is therefore given by the following calculation:

$$t^\circ \text{ Centigrade} = (\tfrac{9}{5}t + 32)^\circ \text{ Fahrenheit},$$

where t° represents the observed temperature.

On the other hand, $t°$ Fahrenheit $- 32 = \frac{5}{9} t$ Centigrade. For temperatures below 0° Fahrenheit, the formula for change to centigrade is —

$$\tfrac{5}{9}(- t° \text{ Fahrenheit} + 32) = - t \text{ Centigrade},$$

and $$\tfrac{9}{5}(- t° \text{ Centigrade}) - 32 = - t \text{ Fahrenheit},$$

and if $\frac{9}{5} - t° = 32$, then the temperature is at 0° Fahrenheit.

In all scientific work the metric system is employed exclusively. If the pupil is not already familiar with this system, he should thoroughly familiarize himself with it before beginning work. The metric system and the centigrade thermometer are used in this book. For a description of a barometer and its use, see Experiment 36.

2. Solution.

a. In a test-tube make a saturated solution of common salt (sodium chloride) in boiling water * by adding more salt than the water will dissolve. Allow the undissolved portion to settle, and pour off the hot, clear, supernatant liquid into a second test-tube, setting the latter aside to cool. (Why cool?) Make a careful examination with the aid of a small magnifier of the crystals which have formed.

Repeat, using blue vitriol (copper sulphate) instead of sodium chloride.

Repeat with potassium chlorate. Use the same amount of water in each of these experiments, and note the difference in the quantities of the three substances which are deposited on cooling.

Repeat the above with a small piece of marble; pour off the water, and evaporate the remainder to dryness in a small porcelain dish over a flame, or, better, on a water bath. (Why use a water bath?) Note if any residue is formed. Is marble soluble in water?

* Distilled water should be used in all chemical work.

***b.* Heat Changes During Solution.**

Take 10 c.c. of distilled water in a test-tube or small beaker, and ascertain the temperature. Add 5 grams of sodium chloride, and stir with the thermometer until solution is effected. Again note the temperature.

Repeat with copper sulphate and with potassium nitrate (saltpetre). (Use a clean beaker and fresh water in each case.)

(The lowering of temperature on solution of salts is caused by the absorption of heat, which is used to produce the fluid condition from the solid.)

3. Water of Crystallization.

a. Carefully dry some crystals of common salt (sodium chloride) by pressing them between filter paper, then place the crystals in a test-tube, and heat in the Bunsen flame.

Repeat with (*b*) crystals of sodium sulphate (Glauber's salt), (*c*) white vitriol (zinc sulphate), (*d*) copper sulphate. What is the difference between the behavior of these substances and that of common salt?

Determination of the water of crystallization.

Put about 2 gms. of copper sulphate crystals in a weighed and clean porcelain crucible, and heat for one hour at 225° in an air bath. Cool in a desiccator and weigh. Heat again and weigh, continuing the operation until the weight is constant. What is the change in appearance of crystals? In weight?

Repeat the same experiment with a second sample of copper sulphate crystals, and then ascertain the per cent of loss in weight which the crystals have suffered in each experiment. Is the result the same in both cases?*

* In performing experiments like the above, care must be taken to have good, clean chemicals. The copper sulphate crystals must be fresh and must not have lost any of their lustre, otherwise no constant results can be obtained.

4. *a*. Efflorescence.

Expose to the air for twenty-four hours some crystals of sodium carbonate on a dry watch-glass. Do the same with some crystals of sodium sulphate. With some crystals of copper sulphate. Note the result in each case.

***b*. Deliquescence.**

Expose to the air, as in the above experiments, some pieces of calcium chloride. Some caustic potash. Some caustic soda. Note the result in each case.

***c*. Heat changes during solution of anhydrous substances capable of uniting with water of crystallization.**

Repeat the experiments under 2 *b*, using 10 cc. of distilled water, and dissolving in this (*a*) 5 grams of fused calcium chloride. (*b*) Repeat with 5 grams of copper sulphate which has been dried at 200°. Note changes of temperature in each case. Compare with the results obtained by dissolving copper sulphate containing water of crystallization.

5. Solution of Gases in Liquids.

a. Under the bell jar of an air-pump, place a beaker of ordinary water and exhaust the air. (Teacher's experiment.)

b. Take a beaker half full of water, place it on a wire netting covering the ring of a retort stand, and heat gently. Note that bubbles of gas pass off before the water comes to a boil.

The following law governs the solution of gases in liquids.

The quantity of a given gas which can be dissolved by a fixed quantity of a liquid varies directly as the pressure on that gas. This form of solution is termed the absorption of a gas by a liquid. Where chemical union takes place between the gas and the solvent, the above law does not hold good. When the pressure is removed, the gas escapes until the solution is saturated at ordinary pressure.

Example: — One cubic centimetre of water dissolves 1.05 cubic centimetres of carbon dioxide at one atmosphere pressure, and 3.15 cubic centimetres at three atmospheres. Ordinary soda-water contains carbon dioxide dissolved under pressure. When the pressure is removed, the gas escapes; i.e., the liquid effervesces.

6. Filtration and Distillation.

The proper arrangement of a filter in a funnel is shown by Fig. 2. The filter paper (unsized paper) is circular and is folded twice, so as to form a quadrant. This is opened so that one thickness of paper is on one side, and three on the other. The paper is now pressed snugly into the funnel, moistened with water, and brought evenly into place by sucking on the stem. A filter should never reach above the edges of the funnel. Take some powdered chalk, suspend it in 20 cc. of water, and pour this on the filter. Note if the liquid passing through is clear. (That which passes through is termed the filtrate, that which remains on the filter the precipitate.)

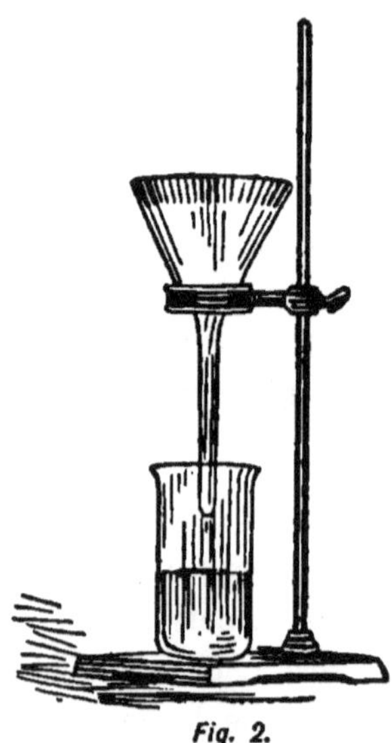

Fig. 2.

Fig. 3 represents the ordinary distilling apparatus, composed of a flask, *D*, with the thermometer, *T*, condenser, *C*, and receiver, *R*. The liquid in the flask is boiled, the vapor is condensed by a current of cold water, which is admitted between the outer jacket and the inner tube of the condenser at the lower opening, and which escapes into a basin at the upper opening.

Distillation.

a. Place in the distilling-flask, which should be of 100 to 200 c.c. capacity, a solution of 2 grams of copper sulphate in 50 c.c. of water. Replace the stopper with the thermometer, and heat until about one-half of the solution has passed over into the receiver. Note the boiling-point of the liquid in the flask. Evaporate the contents of the receiver to dryness, and do the same with the contents of the distilling-flask.

(The contents of the receiver, after distillation, is termed the distillate.)

b. Distil 90 c.c. of ordinary water until about 10 c.c. remain, then add 50 c.c. more, without emptying the distilling-flask, and repeat this operation four times. Now put the contents of the distilling-flask into an evaporating-dish,

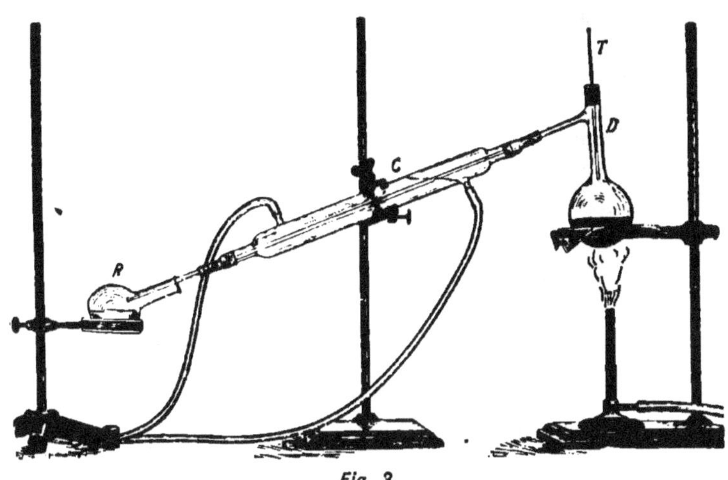

Fig. 3.

and evaporate to dryness on a water-bath. Put the distillate into an evaporating-dish, and also evaporate on a water-bath. What is the difference between the two waters which you have evaporated? What class of substances can be separated from water by distillation?

7. Action of Sodium and Potassium on Water.

a. Take a clean piece of sodium as large as a pea, carefully remove the naphtha with a piece of filter paper, and throw the sodium into a basin of water. Repeat with a piece of potassium. Compare the action of potassium with that of sodium. Be careful to stand aside at least eight feet while the action is going on.

b. **Collecting the gases generated by the action of sodium on water.**

Fig. 4 shows a vessel, preferably of glass or of sheet iron, with a capacity of at least one litre. Nearly fill this vessel with water, and invert over it a small test-tube filled with water. Now wrap in wire gauze a clean piece of sodium as large as a pea, and with a pair of forceps quickly place it under the mouth of the test-tube, or use an inverted wire-gauze spoon, as shown in the cut.* When the tube is full of gas, place the thumb over its mouth, then light a splinter of wood, remove the thumb, and quickly apply the flame to the mouth of the tube. Taste a drop of the water left in the trough. Insert the fingers into the trough, and note the feel of the water.

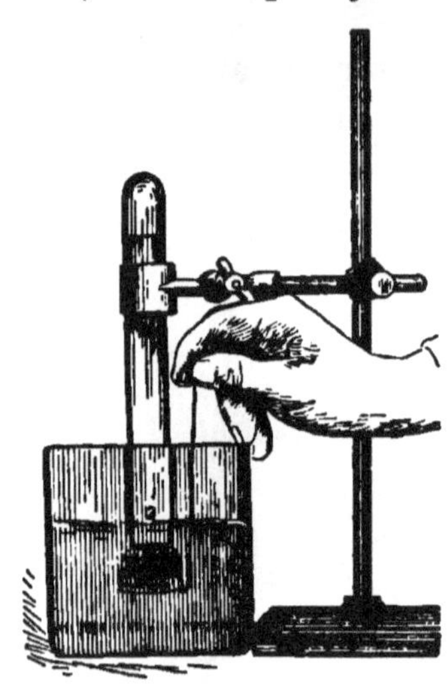

Fig. 4.

8. The Electrolysis of Water.

The apparatus (Fig. 5) consists of a shallow glass vessel with a wide mouth, closed by a water-tight rubber stopper. Through the stopper pass two stout platinum wires about two or three centimetres apart, and extending several cen-

* Care should be taken to test the sodium to be used in any of these experiments by placing a small piece on water, and then standing aside; for unless the metal is clean there is great danger of an explosion. Scraps of sodium which have been kept in the laboratory for some time should never be used. The brown rind must always be carefully removed from the sodium before using.

timetres above and below the stopper. To the lower ends of the platinum wires connect two or more cells of a bichromate battery by means of insulated copper wires. Fill the glass vessel with distilled water, also fill two tubes * with the same liquid, and invert them in the vessel so that their mouths are over the wires; adjust the tubes in a vertical position by means of clamps, and turn on the current. Is there any action? (See page 191.) Now add concentrated sulphuric acid to the water, drop by drop, until there is a satisfactory development of gas at each pole. What gas collects at the negative pole? What gas collects at the positive pole?

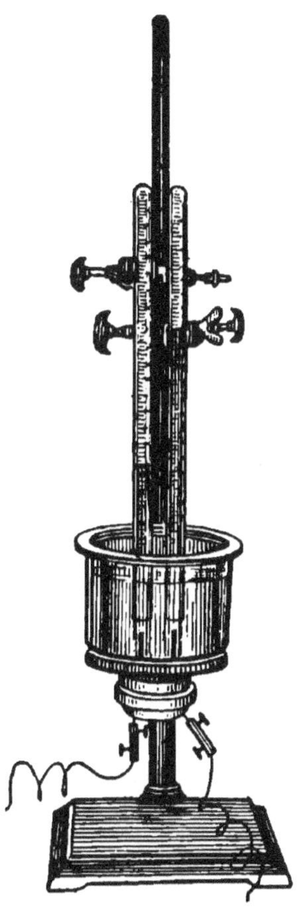

Fig. 5.

When 10 to 15 c.c. of gas has collected at the kathode,† raise the carbon and zinc plates from the battery jar, allow the apparatus to stand for five minutes, and then carefully read off the volumes of the two gases which have collected. Now with the thumb close the tube over the negative electrode, remove it from the basin, invert it, and instantly apply a lighted taper to

* The tubes should be divided into cubic centimetres.

† The negative pole of the battery is the one connected with the zinc, the positive pole is the one connected with the carbon in the cell. The positive pole is termed the anode, the negative pole the kathode, and the liquid which is decomposed is the electrolyte. The two platinum wires are termed the electrodes.

its mouth. Remove the tube over the positive pole in the same way, and insert into it a glowing pine splinter. What is the difference between the two gases? Did you obtain exactly twice as much of one gas as of the other? If this relationship is not exactly correct, what is the reason for the discrepancy?

9. Preparation of Hydrogen and Oxygen for Laboratory Use.

a. The preparation and collecting of hydrogen.

It has been found that hydrogen is liberated by the action of certain substances termed acids (see pages 45 and 57) on certain metals. For example, zinc when covered by dilute sulphuric acid generates hydrogen, while at the same time the zinc is dissolved by the acid, and enters into the formation of a substance termed zinc sulphate. At the present time the means by which it is done does not interest us so much as the study of the properties of the hydrogen which we can obtain.

The apparatus for the preparation of hydrogen is shown by Fig. 6. It consists of a heavy walled flask, into which is fitted a rubber stopper with two holes. Into one of these a so-called safety-tube is fitted (the curved tube with the

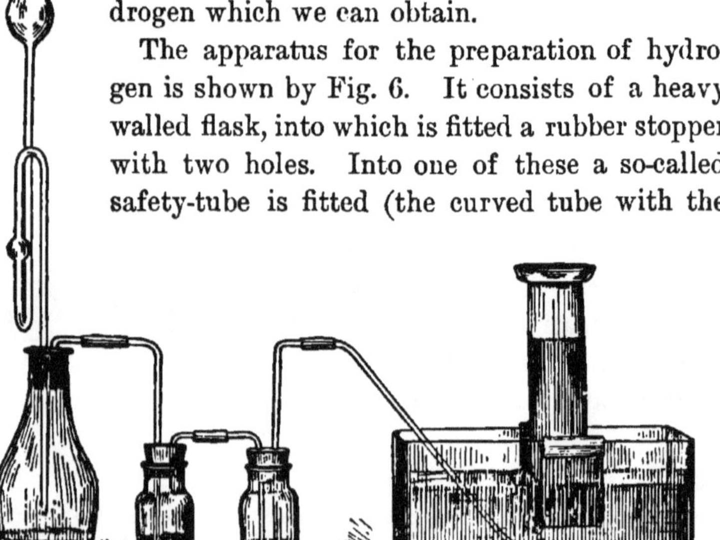

Fig. 6.

funnel top and bulb in the middle; see figure), and into the other is inserted a glass tube bent at a right angle, and reaching just to the bottom of the stopper. When all is ready, place 15 grams of zinc in the flask, insert the stopper, and pour dilute sulphuric acid through the safety-tube, adding more acid from time to time as occasion requires.* To purify the hydrogen generated by zinc and sulphuric acid, pass it through a wash-bottle † containing a solution of two grams of caustic potash in 10 grams of water, and through a second one with concentrated sulphuric acid.

b. **Experiments with hydrogen.**

Fill two or three test-tubes or cylinders with the gas, over water, as shown in the figure. Remove one of these from the water, mouth downward, and quickly insert a lighted taper. Connect an ordinary clay pipe (coated with paraffine on the outside) with the hydrogen generator, and bring the mouth of the pipe into some strong soap-suds, to which has been added a little glycerine or gum arabic. Note if the soap-bubbles rise in the air. Is hydrogen very soluble in water? How would you discover whether a gas is very soluble in water?

c. **The preparation and collecting of oxygen for laboratory use.**

In preparing large quantities of oxygen for laboratory

* Dilute sulphuric acid is prepared by adding one part of commercial acid to six parts of water. In diluting sulphuric acid, pour the acid into the water slowly, but do not pour the water into the acid. Cool the acid by placing the flask in which it is diluted under the hydrant before using.

† A wash-bottle consists of an ordinary wide-mouthed bottle, fitted with a stopper having two holes. Into one hole of the stopper is inserted a tube, bent at a right angle, and reaching to the bottom of the flask. Into the other is fitted a glass tube bent at a right angle, and reaching just to the bottom of the stopper. The gas enters through the former tube, and passes out through the latter. When a liquid is placed in the bottle the gas must bubble through this liquid, and is washed by it. (See Fig. 6.)

use, it is not expedient to decompose water by the electric current. There is a salt-like body (chlorate of potassium) which, when heated to a sufficiently high temperature, gives off *oxygen.*

The apparatus for the preparation of oxygen in large quantities, shown in Fig. 7, consists of a flask of 200 c.c. capacity, fitted with a stopper in which is placed a glass tube bent at two right angles. This tube is put into the flask just to the bottom of the stopper, and its other end is connected with a safety-bottle. This safety-bottle must always be interposed between the water-trough in which the gas is collected, and the generating-flask, in cases where the latter is heated to a high temperature. By this means, if the water should suck back, it is collected in the empty safety-bottle, and a dangerous explosion is avoided. The arrangement of the safety-bottle is shown in the cut. The oxygen passes from the delivery tube of the safety-bottle, and is collected over water just as hydrogen is.

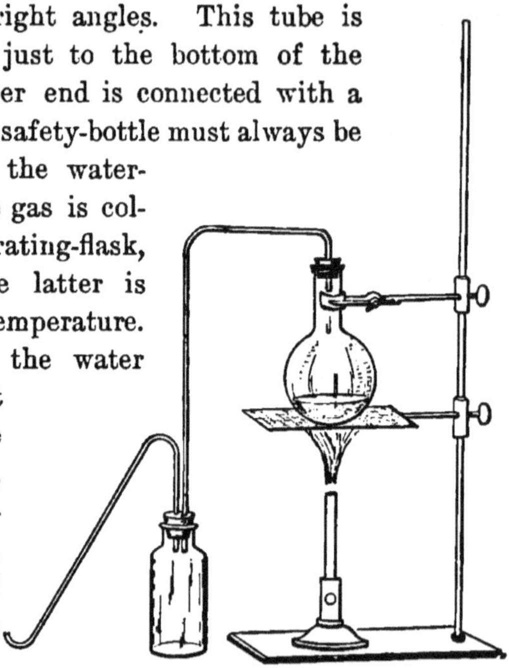

Fig. 7.

In the generating-flask place 15 grams of chlorate of potassium, connect the apparatus, invert several test-tubes in the water-trough, and then heat the chlorate of potassium until gas passes off with moderate rapidity, and until all of the air has been expelled from the apparatus and safety-bottle. Now collect the gas in the test-tubes. Remove one of the tubes, and insert in it a glowing pine splinter. Repeat with a second tube, using a small piece of charcoal heated to redness and placed on the end of an iron wire tightly wrapped around it. Repeat the same experiment with the soap-bubbles which was made with hydrogen. Which gas is more soluble in water, hydrogen or oxygen?

10. Explosion of a Mixture of Hydrogen and Oxygen in the Eudiometer Tube, Fig. 8. (Teacher's Experiment.)

The apparatus in which this explosion is performed is termed a eudiometer tube. This instrument is a glass tube, *A*, closed at one end, and graduated in cubic centimetres. It has two platinum wires inserted near the closed tip, *C*, in such a manner that an electric spark can pass from one to the other. This tube is filled with mercury, and inverted over a mercury trough, *b*. In order to prove the volumetric composition of water, slant the tube to one side, and admit about 10 c.c. of hydrogen prepared and purified as in 9 *a*. Bring the tube to a vertical position, and make careful readings of the following: —

1. Volume of gas.
2. Temperature.
3. Height of barometer.

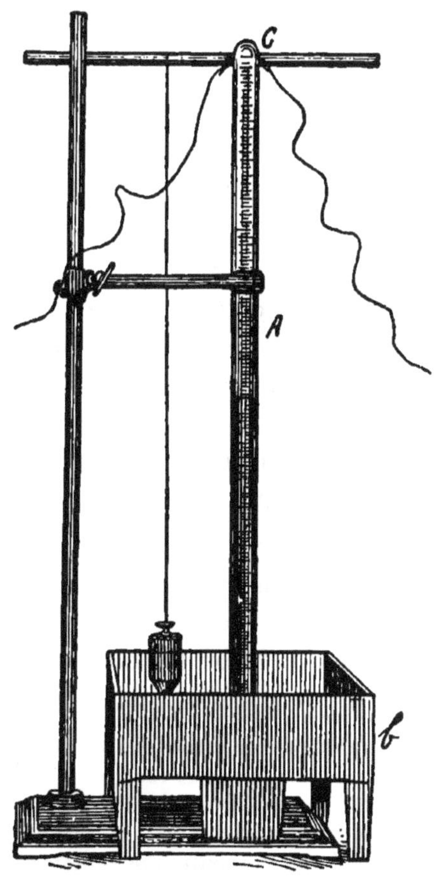

Fig. 8.

4. Height of the column of mercury in the tube (in millimetres).

Place the readings in your note-book. Now again slant your tube, and admit about 7 c.c. of oxygen from the teacher's gasometer, or from the apparatus which was prepared for the generation of oxygen.*

Be careful to run in the oxygen very slowly through a tube drawn down to a small opening, otherwise too much will be admitted by a sudden burst from the generator. Now once more bring the eudiometer tube to a vertical position, and note:—

1. Volume of gas.
2. Temperature.
3. Height of barometer.
4. Height of the column of mercury in the tube.

Clamp the tube tightly in position, with its open end pressed down against a leather washer in the bottom of the mercury trough. Now pass a spark from an induction coil or Leyden jar through the gases by connecting the two poles with the platinum wires in the eudiometer.† After the explosion raise the eudiometer slightly from the washer, and allow the apparatus to stand twenty minutes while cooling. Now note:—

1. Volume of gas.
2. Temperature.
3. Height of the barometer.
4. Height of the column of mercury in the tube.

* Oxygen from a gasometer is pretty sure to contain nitrogen. It is therefore better to prepare fresh oxygen from potassium chlorate. Be sure that the gas has run through the generating-apparatus for a sufficient length of time to expel all air, before collecting in the eudiometer tube.

† Not infrequently the eudiometers come to the laboratory with the ends of the wires so near together that the spark will not be large enough to ignite the gases; if such is the case, force the ends apart carefully with a long glass rod.

Recalculate all the above observations so that the gas volume in each is reduced to 0° and 760° mm. of the barometer.*

If 10 c.c. (corrected) of hydrogen and 7 c.c. of oxygen were admitted to the eudiometer, then 10 c.c. of hydrogen will unite with 5 c.c. of oxygen to form water (2 volumes of hydrogen to 1 of oxygen), and 2 c.c. of oxygen will remain unaltered. The residual gas may be tested for oxygen in the usual manner. It is well to repeat the experiment, this time employing an excess of hydrogen.

11. Proof that Two Volumes of Hydrogen with One of Oxygen form Two Volumes of Water Vapor. Fig. 9. (Teacher's Experiment.)

The apparatus consists, essentially, of a eudiometer surrounded by a glass jacket joined to the eudiometer by means of a tightly fitting triple bored stopper. The three holes in the stopper are for the following purposes: The one in the middle is for the eudiometer tube; the second (*a*), for the admission of steam by means of a glass tube bent with a right angle; and the third for the exit of the condensed water by means of a similarly bent glass tube.

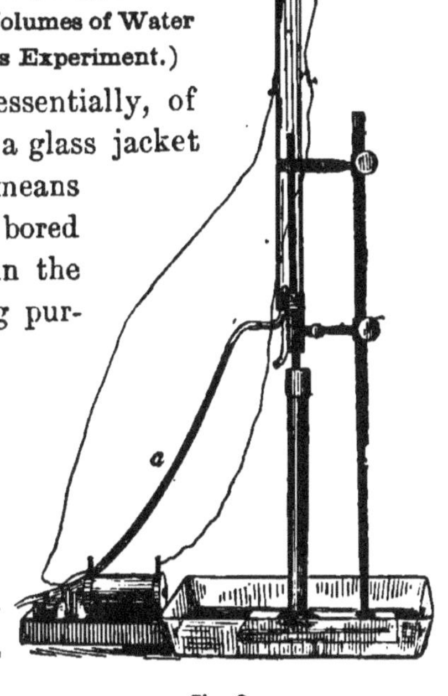

Fig. 9.

* For the formulæ necessary to recalculate gases to standard conditions, and for the reasons for such formulæ, see pages 68, 69, and 70. The experiments numbered 36, 37, and 38 can, if desired by the teacher, be performed in this place before the pupil does the work of recalculating the gas volumes noted in his observations of the above experiment.

Fill the eudiometer with perfectly clean, dry mercury, carefully removing every bubble of air,* and invert it in a cylinder of mercury about 30 centimetres in depth. Now raise the eudiometer until its mouth is just below the surface of the mercury in the trough, and admit about 15 c.c. of the mixture of hydrogen and oxygen obtained from the electrolysis of water.† Now pass a jet of steam into the jacket until the apparatus is heated to 100° C, and note the volume of the gas as well as the height of the column of mercury in the eudiometer tube above that in the trough. Lower the eudiometer tube and jacket so that the open end is well under the mercury, wrap a towel round the top of the mercury trough, and explode the gases as in the previous experiment. Now adjust the tube so that the height of the column of mercury in it is the same as it was before, and make a second reading of the volume of gas, taking care that the tem-

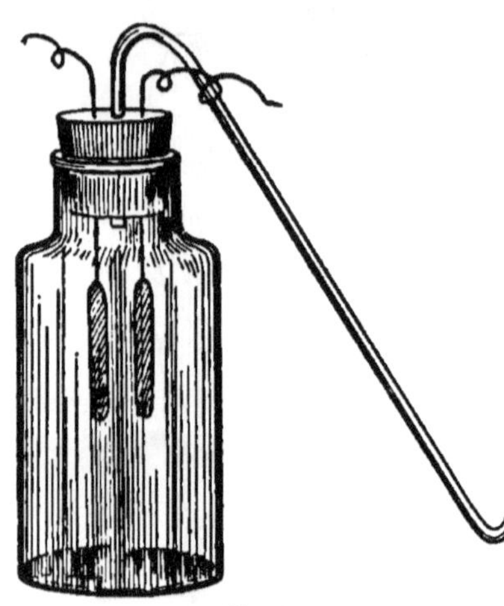
Fig. 10.

* Remove bubbles of air by nearly filling the tube with mercury, closing the open end with the thumb, and slowly passing a large bubble of air back and forth through the entire length of the tube.

† A simple apparatus for generating this mixture of oxygen and hydrogen is shown by Fig. 10. This consists of a wide-mouthed bottle of 100 c.c. capacity, closed with a single bored rubber stopper. In this stopper are inserted two stout platinum wires, hammered flat at the ends, and in the opening is placed a glass tube of one or two millimetres internal diameter, bent as in the figure, and reaching just to the lower edge of the stopper. Almost fill the bottle with dilute sulphuric acid, insert the stopper, connect with the battery, wait for some minutes until the gas has passed off, and then collect in the eudiometer.

perature is still at 100° C. What is the ratio between the volumes of gas before and after explosion? Allow the tube to cool, and observe the remaining volume of gas.

12. Burning of Hydrogen* in Oxygen, and Formation of Water by this Means. (Teacher's Experiment.) Fig. 11.

The apparatus consists of a vessel constricted in the middle, and is fitted with a glass stopcock delivery-tube. A wide globe funnel with a long stem is placed in the upper opening of this vessel. The zinc is placed in the middle globe, and dilute sulphuric acid is added from above until the apparatus is filled to about the middle of the funnel. On opening the stopcock, the

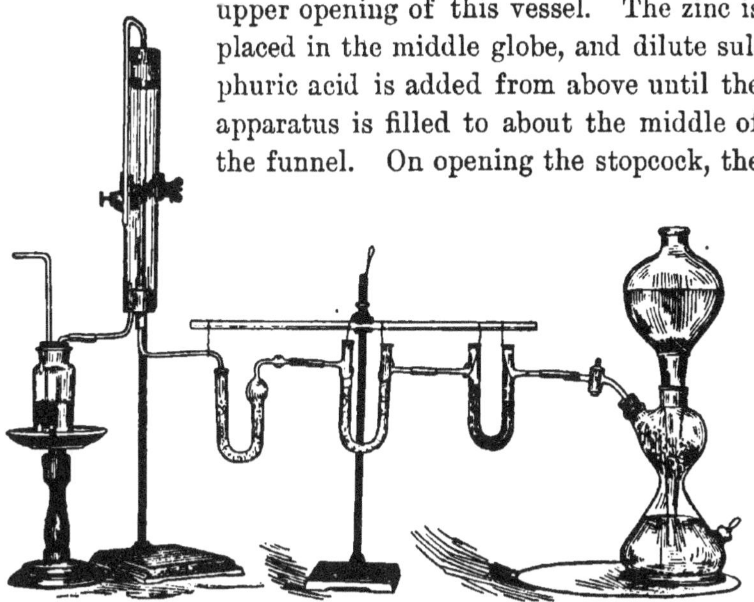

Fig. 11.

acid ascends to the metal. On closing it, the generated hydrogen once more expels the acid from the central globe. In this way the metal can be indefinitely kept out of

* The hydrogen needed must be available in a current which is easily regulated. For this purpose it is best prepared in a so-called Kipp's gas-generator. A description of this generator is given above, and it is shown at the extreme right of the cut.

contact with the acid, and need only be acted on by it when the stopcock is opened.*

A regular current of hydrogen such as is necessary to maintain a small flame, cannot be obtained if the gas is purified by passing through wash-bottles. For this reason so called U-tubes (shown in the centre figure) must be substituted. The contents of these purifying tubes, counting from the generator, are —

1. Pieces of brick moistened with a solution of potassium permanganate.
2. Coarse pieces of solid caustic potash.
3. Granular (anhydrous) calcium chloride.

The exit tube of No. 3 terminates in a platinum tip of a diameter not more than one millimetre, made from a roll of platinum foil, and fused into the open end of the glass tube. The wash-bottle at the extreme left contains concentrated sulphuric acid to dry the oxygen which is passed in from left to right. Both the hydrogen and oxygen which are to be admitted into the vertical cylinder are therefore perfectly dry and pure.

Manipulation: Remove the platinum tipped tube from the vertical cylinder, and start the hydrogen generator. After waiting at least ten minutes, when all the air has been expelled from the train and apparatus, ignite the hydrogen at the tip, and regulate the current so that a flame of not more than one centimetre in length is obtained. In the meantime expel all the air from the vertical cylinder by a steady stream of oxygen from the teacher's gasometer. Now pass the jet of burning hydrogen upward through the central opening in the stopper, so that it is just above the place where the oxygen enters. What collects on the sides of the cylinder? Taste the liquid which soon passes off through the exit tube at the upper end.†

* If necessary the hydrogen can be generated as described in experiment 9 *a*.

† See foot-note on page 25.

13. Examination of the Substance which remains in the Water after it has been acted upon by Sodium.

Cut a piece of sodium as in 7*a*, and throw it into a small evaporating-dish containing 150 c.c. of water. Stand aside until the action is over, and until all the sodium has disappeared. Now add a second piece, and wait as before. Continue the operation until four pieces have been added. Evaporate the solution as far as possible on a water-bath, transfer the small quantity of liquid which is left to a small porcelain crucible, and dry in an air-bath at 150°. Note the appearance of what is left. Has it any resemblance to sodium? Is it soluble in water? Make a very dilute solution of a portion of it, and taste a drop. Compare this taste with that of a very dilute solution of caustic soda which you obtain from the side-table. Compare the feel of the solution with that of a solution of caustic soda.

14. Decomposition of the Caustic Soda obtained by the Action of Sodium on Water. Proof that This Substance contains Hydrogen.

a. A mixture of sodium and caustic soda when heated will develop hydrogen, thereby furnishing absolute proof that only a portion of the hydrogen is expelled from water when it is acted on by sodium; but this experiment probably can be performed successfully only by the teacher. Dry some caustic soda completely by fusing it in an iron crucible over a Bunsen flame. Allow to cool, powder the caustic soda, and place about three grams in a test-tube of so-called infusible glass. Add one-half gram of clean sodium, and fit into the test-tube a single-bored rubber stopper which is provided with a glass tube drawn out to a point, so as to make a burner (Fig. 12). Now gently heat the sodium (holding the test-tube by a holder) until the metal is melted. Gradually increase the

Fig. 12.

temperature, and light the hydrogen at the tip of the burner. This is a conclusive proof that caustic soda contains hydrogen. It has been shown, therefore, that water is first changed by the action of sodium so that a part of the hydrogen is set free, leaving caustic soda; and then sodium, acting on caustic soda, sets free the remainder. The hydrogen in water is therefore divisible into two parts. The experiments which prove that these are two equal parts are too difficult to perform here; it is sufficient to know that .1 gram of sodium, acting on water, sets free 48.22 c.c. of hydrogen; and when the caustic soda formed from this amount of sodium is carefully dried and heated with sodium, a second 48.22 c.c. of hydrogen are liberated. The hydrogen in water is therefore divisible into two equal parts.

15. Inauguration of Chemical Action by Heat. Kindling Temperature.

(Perform these experiments under the hood.)

Place five drops of carbon bisulphide in a dry test-tube, and warm with the hand until the tube is filled with the vapors of the liquid, then insert in the mouth of the test-tube a glass rod which has previously been warmed in the flame of a Bunsen burner.

Cut a piece of phosphorus half the size of a pea (phosphorus must always be cut under water, and handled with forceps), dry it quickly by pressing it between sheets of filter paper, transfer it to a deflagrating spoon, and bring it gradually toward the flame. Repeat with red phosphorus. Sulphur. Carbon.

What is the cause of the explosion in a eudiometer tube when the electric spark is passed from one platinum wire to the other, as in Experiment 10?

16. Preparation of Hydrogen Chloride.

Arrange an apparatus as shown in Fig. 13. The generating-flask should have a capacity of about 500 cubic centimetres. The wash-bottle contains concentrated sulphuric acid.

Place in the generating-flask 25 grams of sodium chloride, pour upon it 50 c.c. of dilute sulphuric acid (2 volumes of concentrated sulphuric acid to 1 volume of water), and heat gently. Collect four small jars of the gas by displacement of air,* covering them with a glass plate as soon as filled. Collect the remainder of the gas in a small beaker of water.

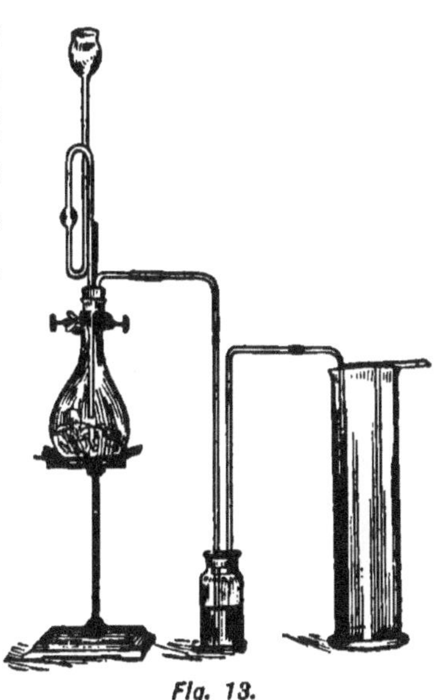

Fig. 13.

17. Properties of Hydrogen Chloride.

Invert one jar of gas in a vessel of water. Taste a drop of the solution. Blow the breath across the mouth of a second jar. Insert into an other jar a burning splinter.

18. *a.* **Preparation of Sodium Amalgam. (Under the Hood.)**

To prepare sodium amalgam place 500 grams of clean, dry mercury in a large clay crucible, and cover with a piece of sheet iron. Cut 5 grams of clean sodium into pieces the size of a hickory nut, and place them in the crucible. To start the reaction, heat a few grams of mercury in a test-tube, slightly raise the cover of the crucible, pour in the mercury, and instantly withdraw the hand. A reac-

* Gases which are specifically heavier than air may be collected by passing the delivery tube, through which the gas is flowing, to the bottom of the jar; gases specifically lighter than air are collected in an inverted jar by passing the delivery tube to the top of the jar. (See Fig. 31.)

tion, accompanied by a flash of light, will occur. Stand aside, so as not to inhale the fumes of the mercury. When the reaction is finished, stir the amalgam with an iron wire, allow it to cool, and place it in a stoppered, wide-mouthed bottle.

b. Decomposition of hydrogen chloride by sodium. Fig. 14.

Take a glass tube 30 cm. in length and 15 mm. in diameter, sealed at one end. Divide this tube by means of two rubber rings slipped over it. One rubber ring is placed at a distance from the sealed end represented by 10 c.c. of the capacity of the tube,* and the other is placed half-way between this ring and the open end of the tube. Have ready a small glass test-tube, *c*, of 10 c.c. capacity, completely filled with sodium amalgam. Now completely fill the large tube with hydrogen chloride, displacing the air, as shown in the cut. The hydrogen chloride must be dried by passing it through a wash-bottle containing sulphuric acid. When the tube is filled, pour in the sodium amalgam, quickly cover the mouth of the tube with the thumb, and bring every part of the gas in contact with the amalgam by repeatedly inverting the tube.

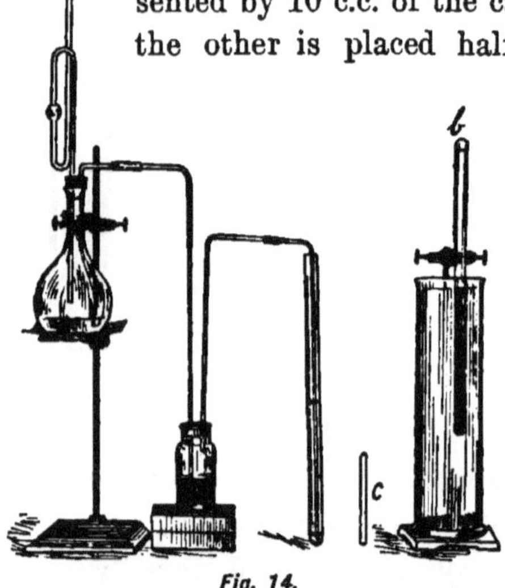

Fig. 14.

* This 10 c.c. of the tube is to be filled with sodium amalgam; hence the volume of hydrogen chloride which will be left, after introducing the sodium amalgam, will be represented by the interval between the ring at 10 c.c. and the remainder of the tube.

Finally, place the open end of the tube in a deep cylinder, under water, and remove the thumb. Lower the tube in the water so that the level within and without is the same (shown by *b* in the figure). What is the relation between the volume of gas remaining and that which was originally present? Now place the thumb over the mouth of the tube, take it from the water, invert it, remove the thumb, and instantly apply a lighted taper. The pupil can convince himself that mercury alone has no action on hydrogen chloride, by repeating the above experiment, using 10 c.c. of pure mercury instead of sodium amalgam. What decomposes the hydrogen chloride?

19. Decomposition of Hydrochloric Acid by Means of the Electric Current. Fig. 15. Hood!

The apparatus consists of a letter-U-shaped tube, which is connected at the centre of the curve with an upright funnel tube, serving for the introduction of the liquid. The two arms terminate in glass stopcocks. The electrodes must be made of the carbon which is used in arc electric lights, because platinum is attacked during the electrolysis of hydrochloric acid. These carbon poles must extend well up into the tubes, and must be protected, except at the tips, by a glass tube slipped over them, and made water-tight by pouring in melted paraffine between the carbon and the tube. This apparatus is designed for teachers' use.

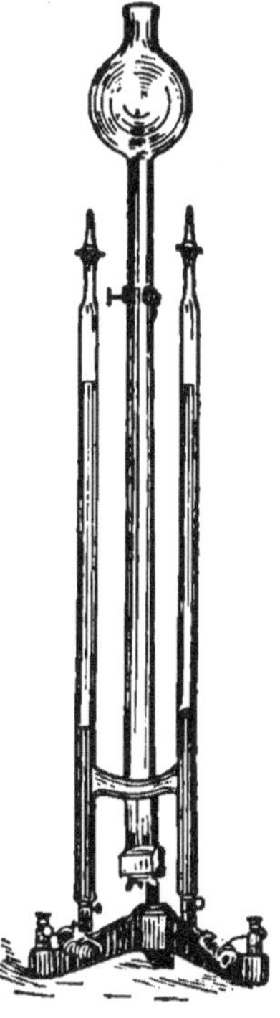

Fig. 15.

A simple apparatus for pupils can be made similar to that shown by Fig. 5, the platinum wires being replaced by two cylinders of carbon, which are connected with the battery by copper wires. Fill either apparatus with a saturated solution of common salt to which has been added one-tenth of its volume of a saturated solution of hydrogen chloride in water. When the current is turned on, a gas will be evolved at the negative pole. When the tube over this electrode is full of gas, remove it, and test for hydrogen. The gas at the positive electrode will not be apparent for some time, because it is soluble, to a certain extent, in the liquid. After a time it, too, will appear. What is its color? Odor? Apply a lighted taper to the mouth of the tube in which this gas is collected. Fill a second tube, and place in it a small piece of moist red calico. If the electric current is allowed to run long enough, equal volumes of the two gases will be evolved in equal lengths of time. Compare this result with that obtained when water is electrolyzed. What does this experiment and the preceding one teach you in regard to the composition of hydrogen chloride by volume?

20. Formation of Hydrogen Chloride from Hydrogen and Chlorine. Volumetric Composition of Hydrogen Chlorine. Fig. 16. Teacher's Experiment.

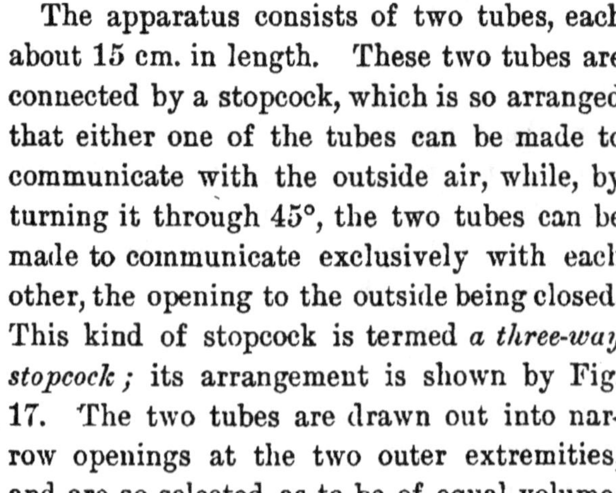

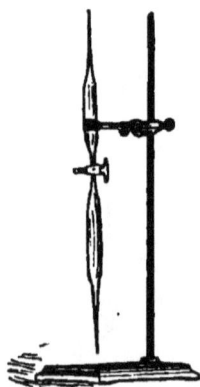

Fig. 16.

The apparatus consists of two tubes, each about 15 cm. in length. These two tubes are connected by a stopcock, which is so arranged that either one of the tubes can be made to communicate with the outside air, while, by turning it through 45°, the two tubes can be made to communicate exclusively with each other, the opening to the outside being closed. This kind of stopcock is termed *a three-way stopcock;* its arrangement is shown by Fig. 17. The two tubes are drawn out into narrow openings at the two outer extremities, and are so selected as to be of equal volume.

Manipulation. Connect the three-way stopcock at *A* with an apparatus furnishing dry chlorine. This is done by the apparatus described in Experiment 21*b*. Turn the stopcock so that the chlorine will pass through one of the tubes, and allow the gas to enter until all air is expelled. Now seal the tip of the tube in a Bunsen flame, and turn the stopcock so that the chlorine tube is closed and the second tube is open to the atmosphere. Fill this tube with dry hydrogen (prepared as in Experiment 9*a*), and when all air is expelled, seal this tip also in the Bunsen flame. Now turn the stopcock so that the two tubes will communicate with each other while they are both shut to the outside air. Allow the apparatus to stand in the daylight for twenty-four hours. Has the color of the chlorine disappeared? Scratch one of the tips with a file, bring it into a vessel containing dry mercury, and break the tip with a pair of pincers. Does the mercury rise in the tube? What does this teach regarding the volume of gas remaining after the experiment, as compared with that before? Hydrogen chloride, as we have learned, is extremely soluble in water, while hydrogen and chlorine are not. Remembering this, we can easily ascertain whether the hydrogen and chlorine have all been used to form hydrogen chloride, by pressing a piece of rubber over the broken tip of the apparatus, and transferring it to a vessel of water.

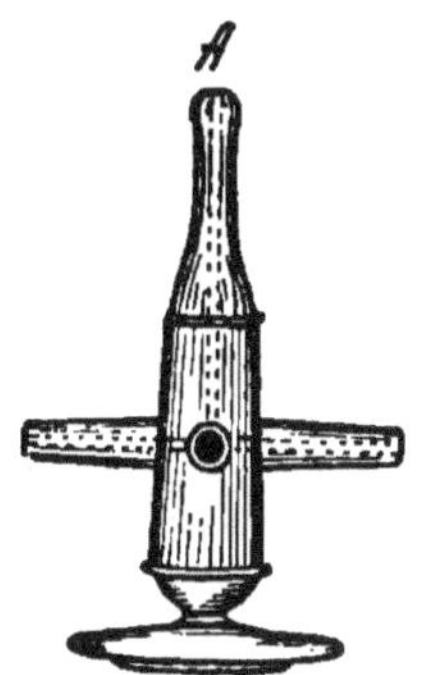

Fig. 17.

If hydrogen chloride has been formed, and if all the hydrogen and all the chlorine have united, the water will rush in and completely fill the apparatus.* Is this the

* Of course, if care has not been taken to expel all the air from the tubes, this air will remain after the experiment, and the result will be inaccurate.

case? What does the experiment teach as regards the relative volumes of hydrogen and chlorine which unite to form hydrogen chloride? As regards the relation between the volumes of hydrogen and chlorine used, and the volume of hydrogen chloride produced? Compare these results with those obtained in Experiments 10 and 11.

21. *a.* Preparation of Chlorine by the Action of Oxygen on Hydrogen Chloride. Fig. 18.

It is not expedient to prepare chlorine for laboratory use by the electrolysis of hydrochloric acid. Chlorine can be prepared by decomposing hydrogen chloride by oxygen, the oxygen uniting with the hydrogen of the hydrogen chloride, while the chlorine is set free. This change takes place only at a high temperature. Prepare an apparatus, *A*,

Fig. 18.

for generating hydrogen chloride as described in Experiment 16, with this difference, that the wash-bottle, *B*, is fitted with a triple-bored rubber stopper. The wash-bottle contains sulphuric acid in order to dry the gases. The three holes are for the following purposes:—

1. For the admission of hydrogen chloride.
2. For the admission of oxygen.
3. For the exit of the mixed gases into an iron tube, *E*, which is placed in a gas furnace (shown in the cut) so that it can be heated to redness.

By this means a mixture of dry oxygen and hydrogen chloride is obtained and passed through the heated tube. The bubbles of hydrogen chloride and of oxygen passing through the wash-bottle should be of equal number in the same interval of time. The ends of the iron tube are fitted with cork stoppers, into which pass the entrance and exit tubes for the gases. The joints are made secure by a coating of plaster of Paris. Between the iron tube and the water-trough beyond is placed a safety bottle, *C*, such as is described in Experiment 9*c*. When the apparatus is ready, the tube is heated to redness in the furnace, and the mixture of hydrogen chloride and oxygen is passed through. Finally pass the gases under a tube, *D*, inverted in a beaker of water. Does the odor of chlorine become apparent? Test the water for chlorine by placing in it a small piece of red calico. Is it bleached?

This experiment shows us that although we can obtain chlorine by the action of oxygen on hydrogen chloride, the method is not practical for laboratory use. However, we need not use gaseous oxygen for this purpose; for it is known that many compounds which contain oxygen give off that oxygen readily under certain circumstances, and that such oxygen, when it is just in the act of being liberated from chemical compounds, readily decomposes hydrogen chloride, setting free chlorine, and forming water. Such an oxygen-containing compound is black oxide of manganese.

b. **Preparation of chlorine from black oxide of manganese and hydrochloric acid. (Experiment under Hood.)**

Arrange an apparatus as shown by Fig. 13, Experiment 16. In the flask place 50 grams of black oxide of manganese, and pour over it through the safety-tube enough of a concentrated solution of hydrogen chloride in water to cover the oxide. In the wash-bottle put sulphuric acid, and collect the gas which passes off by displacement of air, as you did hydrogen chloride. Color of gas? Place in one of the jars a piece of dry colored calico. Place in a second jar a piece of moist colored calico. Odor of gas? (Do not inhale more than a very small quantity of chlorine. Fan the air above the cylinder of chlorine with the hand, and get the odor of the gas while you are at some distance.) Invert one of the jars in a basin of water. Is chlorine soluble in water? As soluble as hydrogen chloride? Pass chlorine into a beaker of water until you have a saturated solution. Color and odor of solution? Does it bleach colored calico?

22. Decomposition of Water by Chlorine in the Sunlight. Fig. 19.

Take a tube 40 cm. long, bent as shown in the figure, and closed at the upper end of the long arm. Completely fill this tube with solution of chlorine in water, invert it as shown in the figure, pour off about 20 c.c. of the solution, and clamp the tube in an upright position. Cover the open end with a test-tube, and leave it in the sunlight for some days. When about 20 c.c. of gas has been collected, close the open end of the tube with the thumb, invert it so that the gas ascends to the bend, then once more bring it to an upright position, allowing the gas to pass into the small arm. Now remove the thumb, and quickly bring a glowing pine splinter into the tube. Taste the liquid remaining in the tube. In this case, what action has the chlorine on the water? Why does chlorine bleach moist colored calico and not dry calico?

Fig. 19.

23. Action of Hydrochloric Acid on Metals.

a. In a test-tube place a little sodium amalgam, and pour over it some hydrochloric acid. Apply a lighted taper to the gas which passes off, or, better, arrange the test-tube as shown by Fig. 12, and light at the tip. After the action has ceased, pour off the liquid from the mercury, and evaporate to crystallization on a water-bath. Examine the crystals under a magnifying-glass. In what experiment have you seen similar crystals? What substance is formed?

b. In a test-tube place some mercury, and pour on it hydrochloric acid. Is there any action? In the above experiment, what acted on the hydrochloric acid?

c. Repeat *a*, using a few pieces of granulated zinc instead of sodium amalgam. If necessary, filter the solution which is formed, and evaporate.

d. Repeat *a* with iron filings.

e. Repeat *a* with a piece of magnesium wire.

24. Action of Hydrochloric Acid on the Oxides of Metals.

a. In a test-tube place a small piece of quicklime (calcium oxide). Pour on it dilute hydrochloric acid. Evaporate the solution, and examine as in 22*a*.

b. Repeat, using zinc oxide.

c. Repeat, using oxide of iron.

d. Repeat, using magnesium oxide. In the last three cases determine whether water alone has any solvent action.

25. Action of Hydrochloric Acid on the Hydroxides of Metals.

a. In a test-tube place a small piece of caustic soda (sodium hydroxide), dissolve it in water, and place a drop, taken out with a glass rod, on a piece of paper dyed with red litmus. Now add dilute hydrochloric acid to the liquid in the test-tube until a drop turns a strip of paper dyed with blue litmus to red.* Now evaporate as in 22*a*, and examine the crystals.

* Acids turn blue litmus (a vegetable dye) to red; alkalies, like caustic soda, turn red litmus to blue.

b. In a test-tube place a small piece of quicklime (calcium oxide), and add a few drops of water. Wait a few minutes. Pour on 10 c.c. of water, shake well, and filter. Take a drop of the filtrate, and put on a piece of red litmus paper. Now add hydrochloric acid, and proceed as in *a.*

c. Obtain some magnesium hydroxide, and add hydrochloric acid until solution is effected. Evaporate as in *a.*

d. Repeat, using zinc hydroxide.

e. Repeat, using hydroxide of iron.

26. Use of Indicators.

Solutions which contain alkalies (sodium hydroxide, potassium hydroxide, calcium hydroxide) are termed alkaline. The reverse of an alkaline solution is an acid solution. We are acquainted with a number of colored substances which assume colors varying with the solution, whether it be alkaline or acid. Such substances are termed indicators. Litmus (Experiment 25), as we have seen, is blue in the presence of alkalies, and red in the presence of acids. It is obvious that, with the aid of litmus, we can tell whether a solution is acid or alkaline; but there are certain objections to this vegetable dye. These objections are that it is changed in color not only by the alkaline hydroxides, but by a number of salts as well. For this reason litmus has largely been superseded by a number of other indicators, the most convenient of which is a dye made from coal-tar, and called methyl orange.

Dissolve one-half a gram of methyl orange in 25 c.c. of water. To a little of this solution in a test-tube add two or three drops of hydrochloric acid. To this acid solution add a dilute solution of caustic soda until the color changes. At the point where the color changes, the solution is said to be neutral; i.e., there is neither hydrochloric acid nor sodium hydroxide present, for the acid has neutralized the hydroxide to form sodium chloride and water. Repeat, using potassium hydroxide. Repeat, using a solution of calcium hydroxide.

27. Neutralization of Known Quantities of an Acid by a Base, and vice versa.

Liquids are most accurately measured by means of burettes. (Fig. 20.) A burette is a glass tube graduated into $\frac{1}{10}$ c.c., open above, closed below either with a rubber tube closed with a pinchcock and terminating in a glass tip drawn to a point, or with a glass stopcock. By opening the stopcock, liquid runs out, and its amount can be accurately measured by the graduation above.

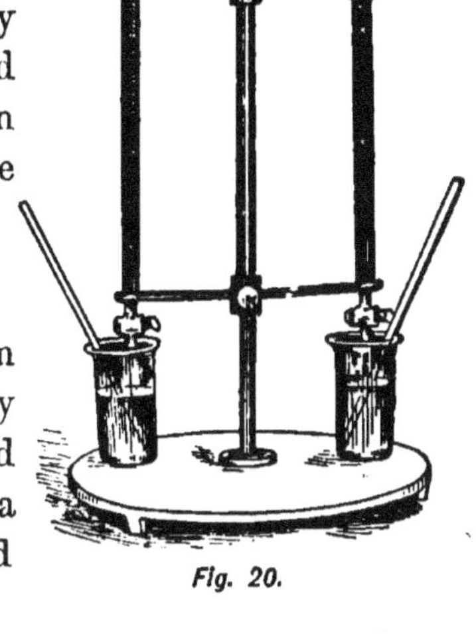

Fig. 20.

a. **To make a solution of alkalies.**

1. Dissolve 4 grams of sodium hydroxide, weighed as accurately as possible, in 500 c.c. of distilled water. Place the solution in a tightly stoppered bottle, and keep for future use.

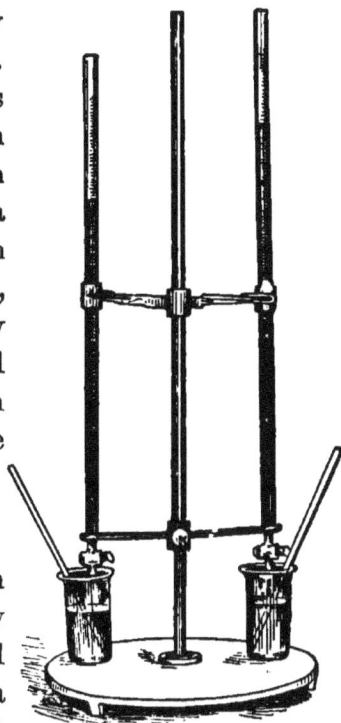

2. Make and bottle a solution of the same quantity of potassium hydroxide in the same amount of water.

b. **Solution of hydrochloric acid.**

Measure accurately 5 c.c. of hydrochloric acid (1 part of pure concentrated hydrochloric acid to 1 part of water), and dilute to 250 c.c. Preserve this solution for future use in a stoppered bottle.

Have two burettes; fill one with solution *a* 1, the second with solution *b*, and cover the open ends by slipping over them two inverted test-tubes. Clamp the burettes in an upright position, as shown in the cut.

In a clean beaker place 10 c.c. of solution *b*, add a few drops of a solution of methyl orange (Experiment 26), and, after noting accurately the quantity of solution in the burette containing solution *a* 1, add this, drop by drop, to the acid in the beaker, at the same time stirring with a glass rod. Stop adding solution *a* 1, at the point where the color of the methyl orange just changes permanently, and note the number of cubic centimetres of the alkali which have been used. What amount of sodium hydroxide, in grams, does this represent? Repeat with a second 10 c.c. of acid, and note whether the two results agree. If they do, pour out the alkali solution from the burette, wash the instrument carefully with distilled water, dry it, and fill with solution *a* 2. Take 10 c.c. of the acid solution, and proceed as before. What amount in grams of potassium hydroxide was used to neutralize 10 c.c. of the acid? Is this amount the same as the quantity of sodium hydroxide used to neutralize the same quantity of acid? If it is not, what is the ratio between the amounts of sodium hydroxide and potassium hydroxide which are used to neutralize the same quantity of acid?

c. Preparation of a solution containing a known quantity of hydrochloric acid.

Into a distilling-flask, arranged as in Fig. 3, place 15 grams of sodium chloride; add a cold mixture of 40 grams of sulphuric acid and 80 grams of water. Arrange the condenser so that it terminates in an ordinary pair of wash-bottles (shown by Fig. 6), each containing about 10 c.c. of water. Place the thermometer tightly in the distilling-flask, and distil until one-half of the liquid has passed over. Allow it to cool, add another 80 grams of water to the distilling-flask, and distil again. All the hydrochloric acid will now have passed over with the distillate. Take the contents of the wash-bottles, carefully wash them into a graduated cylinder, and dilute to 500 c.c. Fifteen grams of sodium chloride produce 9.18 grams of hydrochloric acid, therefore the 500 c.c. of solution contain 9.18 grams of hydrochloric acid. Take 10 c.c. of this solution, and neutralize with solution *a* 1. With solution *a* 2. How much by weight of sodium hydroxide is neutralized by 1 gram of hydrochloric acid? How much of potassium hydroxide? What is the ratio between the amounts of sodium hydroxide and of potassium hydroxide which will neutralize 1 gram of hydrochloric acid? Compare this result with those obtained in this Experiment, *b*. The quantities of sodium hydroxide and of potassium hydroxide which will exactly neutralize the same amount of hydrochloric acid are said to be equivalent.

28. *a.* **Burning of Sulphur in Oxygen and the Product formed.**

Fill a 32-oz. wide-mouthed bottle with oxygen. Place about .5 grams of sulphur in a deflagrating-spoon passed through a piece of sheet tin large enough to cover the mouth of the bottle. Ignite the sulphur in the flame of a burner, and lower it into the bottle. When combustion is completed, remove the spoon, cover the jar with a glass plate, and allow it to stand for a few minutes. Color of gas? Lower a burning pine splinter into the jar. Moisten a piece of blue litmus paper, attach it to the glass cover, and allow it to stand in the bottle. Changes of color? Test the odor of the gas as you did with chlorine. (Experiment 21*b*.) The product of this combustion is termed sulphur dioxide.

b. Preparation of sulphur dioxide for laboratory use.

It is not expedient to prepare sulphur dioxide for laboratory use by the above method. In order to obtain sulphur dioxide in large quantities, advantage is taken of the following fact: —

Sulphuric acid is composed of hydrogen, sulphur, and oxygen. When it is heated with certain metals it gives up a portion of this oxygen, much in the same way as black oxide of manganese gives up its oxygen to hydrochloric acid (see Experiment 21*b*). What is left of the sulphuric acid after it has lost a portion of its oxygen then breaks down into sulphur dioxide and water, the sulphur dioxide being a gas. The best metal to use for this purpose is copper.

Arrange an apparatus as in Experiment 16, Fig. 13. In the generating-flask place 25 grams of copper shavings, and add 100 c.c. of concentrated sulphuric acid through the safety-tube. Connect all parts of the apparatus (the wash-bottle should contain concentrated sulphuric acid), and heat by means of a Bunsen burner until gas begins to come off. At this point lower the flame until you obtain a regular and slow evolution of gas. Collect the evolved gas by displacement of air in four small jars, which you cover with glass plates as soon as filled, and pass the excess of gas into a beaker of water. Odor of gas? Compare with odor of gas obtained from burning sulphur. Invert one of the jars in a vessel of water. Is the gas very soluble in water? Lower a lighted splinter into another jar. Pour into one of the jars a little blue litmus solution, and allow it to stand for a short time. Obtain a red rose, dip it in water, and suspend it in a jar of sulphur dioxide for some time. A solution of sulphur dioxide is easily changed back to sulphuric acid by means of oxygen. To a few drops of sulphur dioxide solution, in a test-tube, add, drop by drop, a solution of chlorine in water until the odor of sulphur dioxide disappears. What is the means by which chlorine is able to add oxygen to sulphur dioxide?

29. Changing of Sulphur Dioxide to Sulphur Trioxide by Means of Oxygen.

Arrange an apparatus as shown in Fig. 21. The generating-flask is the same as in Experiment 28; but the wash-bottle contains a triple-bored rubber stopper, arranged as in Experiment 21, Fig. 18. Into one of the holes of the wash-bottle pass a tube connected with the sulphur dioxide generator; into the second pass a

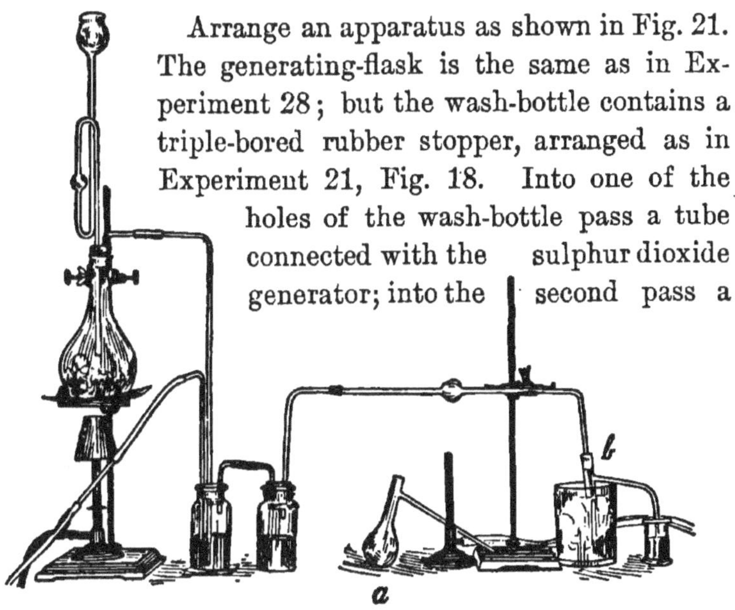

Fig. 21.

tube from the oxygen gasometer. By this means a mixture of sulphur dioxide and oxygen is obtained in the wash-bottle. This mixture passes from the third tube into a second wash-bottle and then into a tube of infusible glass, in the middle of which is blown a bulb, into which is stuffed, with a glass rod, a little platinized asbestos.* The farther end of this bulb tube is connected with a small distilling-flask,

* Platinized asbestos is asbestos coated with a very fine deposit of platinum. It has been found that many chemical reactions between gases take place much more readily if they are passed over heated, finely divided platinum, such as is deposited on platinized asbestos. The reason for this is that the platinum condenses the gases on its surface, and gives them a more intimate contact. Platinized asbestos is furnished by dealers in chemicals.

b, which is surrounded by snow or ice, and the exit tube of which dips under mercury.* The mercury constitutes a valve for the escape of gases, while at the same time it prevents moisture from getting back into the apparatus. When all is ready, pass the mixture of sulphur dioxide and oxygen slowly through the bulb tube, and heat the platinized asbestos by a Bunsen burner. The sulphur trioxide which is formed is condensed in the small receiving-flask. After the apparatus has been running about half an hour, stop the inflow of gases, and remove the receiving-flask. Is sulphur trioxide a solid or a liquid? If a solid, warm the flask slightly with the hand. Does the sulphur trioxide melt?

30. Formation of Sulphuric Acid from Sulphur Trioxide and Water.

Surround a receiver containing sulphur trioxide with cold water, and carefully add water, drop by drop, until all is liquid. Dilute a few drops of this liquid with a good deal of water in a test-tube. Taste a drop. Add a few drops to a solution of methyl orange. Place a drop on a strip of blue litmus paper. Take a few drops of sulphuric acid from your reagent bottle, dilute as above, and repeat these tests. Evaporate any excess of water in the sulphur trioxide by placing the solution in a small evaporating-dish on a water-bath. Compare the appearance of the remainder after evaporation with ordinary sulphuric acid, and preserve it until the next experiment is finished.

31. Dehydrating Action of Sulphuric Acid.

a. Put a little sugar in a test-tube, and pour on it just enough concentrated sulphuric acid to cover. Set it aside in the test-tube rack. Put some concentrated sulphuric acid in a test-tube, place in it a small piece of wood, and allow it to stand. The effect on the wood and on the

* The shape and relative size of the distilling-flask is shown by *a* in the figure.

sugar is due to the fact that sulphuric acid has a great tendency to take up water. Sugar and the greater portion of the wood contain hydrogen and oxygen in exactly the proportions necessary to form water. When acted on by sulphuric acid a reaction takes place, and the water is absorbed by the acid. Repeat the above, using the sulphuric acid formed by the action of water on sulphur trioxide.

b. **Changes of temperature when sulphuric acid is diluted.**

Repeat Experiment 4*c*, gradually adding concentrated sulphuric acid to water, and stirring it with the thermometer. Note changes in temperature. Place some concentrated sulphuric acid in a test-tube, and surround with a mixture of snow and salt. Does the acid freeze? Weigh a small stoppered flask, then add 10 c.c. of sulphuric acid, and again weigh; pour out the acid, and wash well with water. Now place 10 c.c. of water in the flask, and weigh. What is the specific gravity of the acid?

32. Action of Dilute Sulphuric Acid on Metals.

Repeat Experiment 23*a*, *b*, *c*, *d*, and *e*, using dilute sulphuric acid instead of hydrochloric acid (1 part concentrated acid to 20 parts of water). What substances are formed? What gas passes off?

33. Action of Sulphuric Acid on the Oxides of the Metals.

Repeat Experiment 24*a*, *b*, *c*, *d*, using dilute sulphuric acid instead of hydrochloric acid. What substances are formed?

34. Neutralization of Sulphuric Acid by Potassium Hydroxide and Sodium Hydroxide.

Use burettes, as in Exp. 27, Fig. 20. Measure accurately 10 c.c. of sulphuric acid (containing 1 volume of pure, concentrated sulphuric acid, specific gravity 1.84, to

6 volumes of water*), and dilute this to 500 c.c. Put it in a stoppered bottle, and preserve it.

a. Place 50 c.c. of this solution in a beaker, add a few drops of methyl orange solution (Experiment 26), and run in from a burette a solution of sodium hydroxide prepared exactly as in Experiment 27*a*, 1. After about 30 c.c. have been added, begin to add your alkali cautiously, drop by drop, until the solution becomes neutral. Evaporate the solution on a water-bath until crystals separate. Examine these carefully under a magnifying-glass. Dry some of the crystals with filter paper, place them in a test-tube, and heat cautiously. Compare the result with that obtained in Experiment 3*b*. Finally heat the crystals to the full heat of a Bunsen burner, and test the water which has collected on the upper part of the test-tube with blue litmus paper.

b. Repeat *a*, using a solution of potassium hydroxide prepared as in Experiment 27*a*, 2.

Are the quantities of sodium hydroxide and potassium hydroxide necessary to neutralize the same quantity of acid equal? If not, what is the ratio between the two? Is this ratio the same as that obtained in Experiment 27 in neutralizing hydrochloric acid?

c. Repeat *a*, placing in the beaker 100 c.c. of the acid, and adding just as much sodium hydroxide as was found necessary to completely neutralize 50 c.c. Evaporate the solution on the water-bath, examine the crystals, and heat as in *a*. Test the moisture which collects on the sides of the test-tube with blue litmus paper. Is it acid? What is the difference between the salt formed in this experiment and the one in *a*? If you had used the same amount of sulphuric acid in each case, what would be the

* A large quantity of this acid can be prepared by the teacher, and kept for use.

ratio between the amount of sodium hydroxide used in *a* and in *c*.

d. Repeat *c*, with a solution of potassium hydroxide prepared as in Experiment 27*a*, 2.

35. Comparison of the Results obtained with Sulphuric Acid and Those with Hydrochloric Acid.

If the sulphuric acid which you used was accurately prepared, 10 c.c. contained .0511 gram of sulphuric acid. How much sodium hydroxide by weight was necessary to neutralize completely one gram of sulphuric acid? How much potassium hydroxide? How much sodium hydroxide reacted with one gram of sulphuric acid in *c*? In *d*? What is the ratio between the quantities of sodium hydroxide and potassium hydroxide necessary to neutralize one gram of sulphuric acid? Is this ratio the same as that between the sodium hydroxide and potassium hydroxide

Base.	Hydrochloric Acid.		Sulphuric Acid.		Weight of Hydroxide in the Solution which was used.	
	10 c.c.	20 c.c.	10 c.c.	20 c.c.	for 10 c.c. of Sulphuric Acid.	for 10 c.c. of Hydrochloric Acid.
	c.c.	c.c.	c.c.	c.c.	grams.	grams.
Sodium hydroxide.						
Potassium hydroxide.						

	Weight of Hydroxide necessary to neutralize one Gram of Hydrochloric Acid.	Sulphuric Acid.
Sodium hydroxide.		
Potassium hydroxide.		

Ratio, —
sodium hydroxide:
potassium hydroxide : 1 : x.

Ratio, —
sodium hydroxide:
potassium hydroxide : 1 : x.

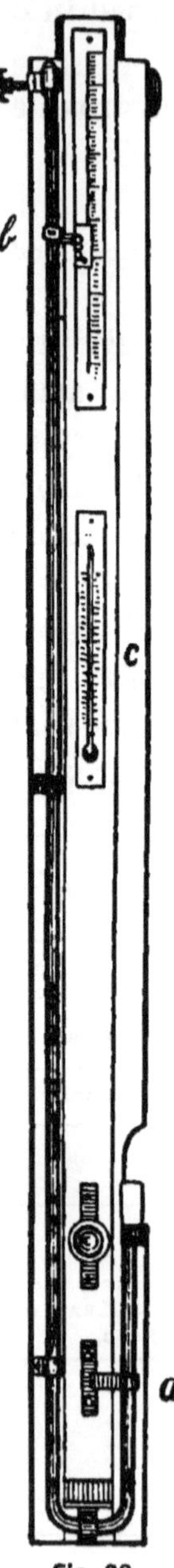

Fig. 22.

necessary to neutralize one gram of hydrochloric acid? Concordant results can be obtained in the above experiments *only* by the most careful work. Each determination should be made at least twice, so as to insure accuracy. The table on preceding page will be found expedient in tabulating results.

36. Construction of a Barometer.

a. Seal one end of a glass tube one metre in length and about one cm. in internal diameter. Nearly fill this tube with dry mercury, close with the thumb, and remove small air-bubbles by passing a larger bubble to and fro in the tube. Now entirely fill the tube, again close, and invert in a vessel of mercury. Clamp the tube in an upright position, and measure with a metre-stick the distance between the level of the mercury in the lower vessel and the level in the tube.

b. The barometer shown by Fig. 22 is constructed on the following principle. The tube is curved so that the shorter arm, *a*, takes the place of the lower vessel described above, the mercury being held in place, with its lower level at *a*, by the pressure of the atmosphere. The tube is filled with mercury, carefully heated until all small air-bubbles are expelled, and then inverted. At the upper level of the mercury, and extending 150 mm. above and below it, is placed a portion of a metre scale, the 0 point of which is given by a movable mark at *a*. The scale has attached to it a sliding vernier, *b*, so that the level of the mercury can be read to $\frac{1}{10}$ of a millimetre. As the height of the barometer varies, so must the level of the mercury above and below vary. However, as the measurement must always begin at the same point, provision is made for adjusting the level of the lower mark so that it corresponds to the lower level of the mercury. When the vernier is placed at the upper level, the corresponding mark on the scale gives the total height of the barometer. *c* represents a thermometer attached to the instrument.

c. As water is specifically much lighter than mercury, it must be true that the atmosphere can support a higher column of water than of mercury. In order to prove this, cover the mercury vessel in this Experiment, *a*, with about an inch of water, and then carefully raise the barometer tube until its mouth is just above the level of the mercury, but below the level of the water. The mercury flows out, the water takes its place, and completely fills the tube.

37. Proof That a Given Volume of Gas varies Inversely as the Pressure on It.

The apparatus is arranged as in Fig. 23, 1. It consists of a tube about one and one-third metres in length and 20 cm. internal diameter. It is open at both ends, and curved so as to form a short arm of about 30 cm. in length. The short arm is closed with a rubber stopper, single-bored, into which is fitted an absolutely tight glass stopcock. The tube is fastened to a board, on which are marks indicating distance, in inches or centimetres.

Manipulation: Open the glass stopcock, and carefully pour in clean, dry mercury through the long arm, until the level in both arms stands at 0 (Fig. 23, 1). Now close the stopcock. Note the volume of confined air, which obviously is under a pressure of one atmosphere. Note also the height of the barometer. Again pour mercury into the long arm, until the difference in the level of the mercury is equal to one-half the observed barometric height (Fig. 23, 2). The confined air is now subjected to a pressure of one and one-half atmospheres. What is the ratio between the present volume and the original volume of air? Again add mercury until the gas is subjected to a pressure of two atmospheres (Fig. 23, 3). What is now the volume of gas? What is the ratio between this and the original vol-

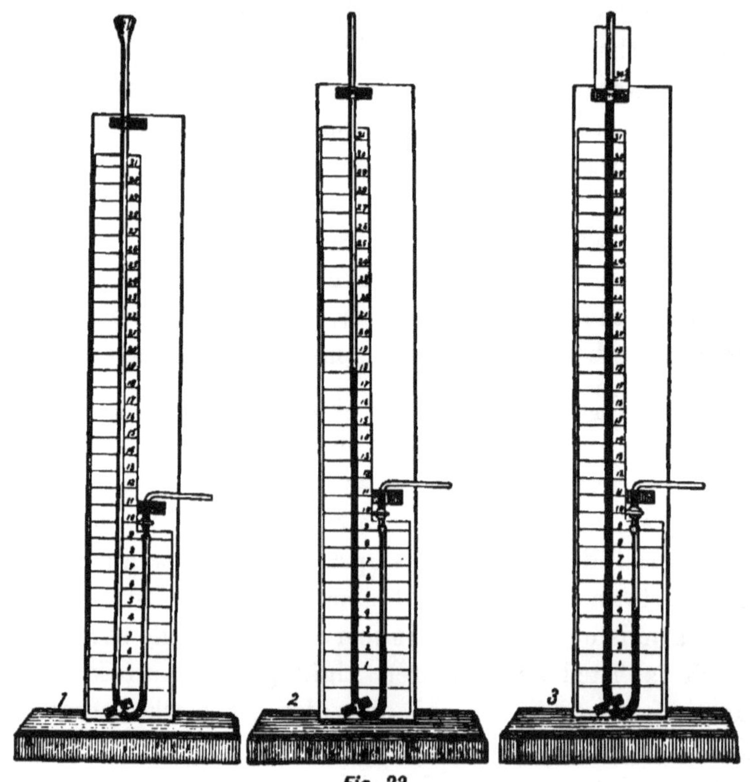

Fig. 23.

ume? Of course, if a sufficiently long tube is taken, this experiment can be extended so as to include 3 or 4 atmospheres, and it can be repeated with other gases (hydrochloric acid, sulphur dioxide, oxygen, and hydrogen). Other gases can be filled in by attaching the gas generator to the open stopcock, while at the same time enough mercury is placed in the tube to bring both sides to 0. The gas will then be forced in, will bubble out, and through the mercury, and expel the air.

38. Influence of Vapor Pressure on the Volume of Gas enclosed in a Barometric Tube.

a. Arrange a barometer as in Experiment 36 *a*, marking the height of the mercury column with a rubber ring. Slip under the open mouth of the barometer tube the end of a curved pipette (Fig. 24) filled with water, the upper end of which you have closed with the finger. Remove the finger, and blow into the upper end of the pipette, so as to force about 10 drops of water into the barometer. What is the level of the mercury after introducing the water? Note the temperature, and measure in millimetres the difference between the original height of mercury and the present. Warm the tube at its upper end, where the water is, by holding the hand round it for a few minutes. Warm it in a cloth dipped in hot water. Note change in the height of the barometer in each case.

Fig. 24.

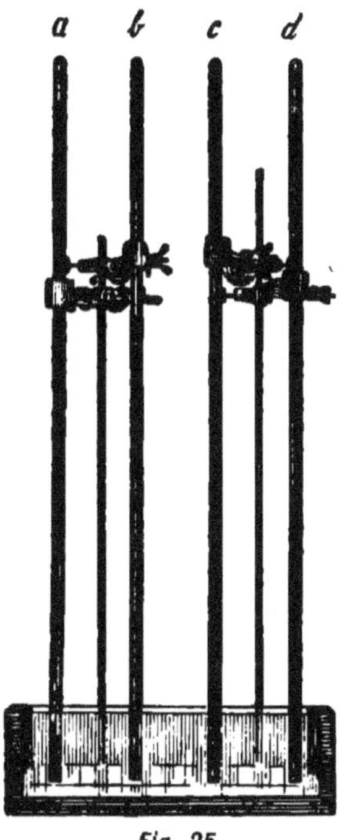

Fig. 25.

b. Repeat *a*, using alcohol instead of water.

c. Repeat *a*, using ether instead of alcohol.

Fig. 25 represents four barometer tubes. Tube *a* is an ordinary barometer; tube *b* contains water vapor; tube *c*, alcohol; and tube *d*, ether.

d. **Expansion of gases by heat.**

Repeat Experiment 36*a*, with the difference that in filling the barometer tube you allow 5 c.c. of air to remain. Mark the height of mercury in the tube by a rubber ring. Now warm the enclosed gas with the hand. With a cloth dipped in hot water. To perform this experiment accurately, so as to ascertain the degree of expansion with each degree of temperature, is too elaborate for this work. The pupil must remember, arbitrarily, that gases expand $\frac{1}{273}$ for each degree of increase in temperature. Using the formula on page 70, ascertain what the volume of 25 c.c. of air would be at 760 mm. pressure, if this air is measured at 743 mm. What is the volume of 25 c.c. of air at 0° and 760 mm. if the air is measured at 22° and 760 mm.? Twenty-five c.c. of air, containing water vapor, and enclosed in a tube over mercury, are under the following conditions:—

Temperature = 30°.
Barometer = 725 mm.
Height of mercury column = 425 mm.

What is the volume of the dry gas at 0° and 760 mm.?

e. Fill a glass tube of 100 c.c. capacity, graduated in half c.c., one-fourth full of mercury, and invert it in a cylinder of mercury about 30 c.m. in depth. Lower the tube so that the level of the mercury without and within is the same. Under what pressure is the enclosed volume of gas? Note accurately the enclosed volume of gas and the height of the barometer. Raise the tube a few centimetres, clamp it in this position, and allow it to stand for five minutes. Again read the volume of gas, and measure carefully the height of the column of mercury in the tube above that in the cylinder.

Repeat, raising the tube still higher.

If v = the first volume,
v_1 = the second volume,
B = barometric height,
h = height of column of mercury in tube,

show that $v : v_1 :: B - h : B$.

39. Preparation of Nitrogen from the Atmosphere. Fig. 26.

The apparatus consists of a tube of infusible glass, with a bulb blown in the middle. This bulb is filled with copper filings. At the right this tube is connected by means of a capillary glass tube with a large bottle fitted with a double-bored stopper, into one hole of which a siphon is tightly

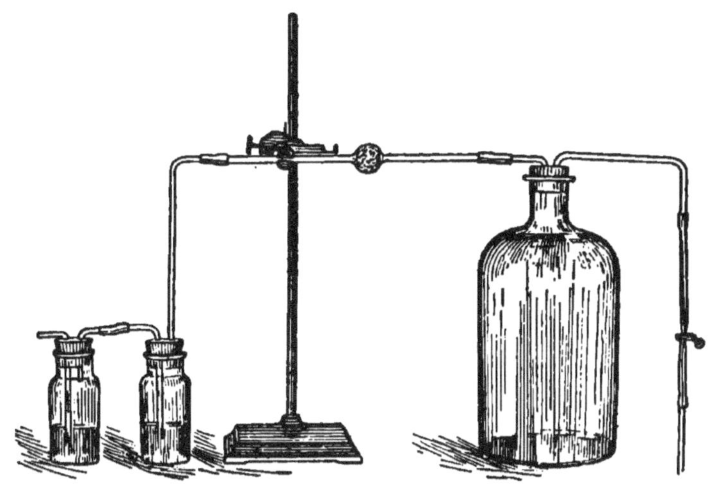

Fig. 26.

fitted. The siphon should be fitted with a rubber tube and pinchcock, so as to regulate the flow. The end of the hard glass tube farthest from the siphon is connected with two wash-bottles, both of which contain strong sodium hydroxide solution. These wash-bottles are to remove carbon dioxide from the air, which is to be drawn through the tube containing the copper. When all connections are made air-tight, heat the copper filings with a Bunsen burner, and after a few minutes start the aspirator slowly. Air will now be drawn into the apparatus through the wash-bottles and over the copper filings. That which remains will be collected in the large bottle. Change of color will occur in copper. Why? After the bottle is one-half full of gas, tightly close the siphon, disconnect the apparatus, raise the double-bored stopper, and insert a burning pine splinter. Note the difference of the behavior of the gas as compared with ordinary air.

Pass a current of air over heated copper filings until all original copper color has disappeared. Connect the tube containing this altered copper with a hydrogen generator, delivering pure dry hydrogen (Experiment 12, Fig. 11), and after the gas has passed over for at least five minutes, gently heat the altered copper by means of a Bunsen burner. What is the change in appearance. What collects in the farther end of the tube? What chemical change has taken place?

40. Combustion in Oxygen.

Prepare an oxygen generator as in Experiment 9*c*. Collect the gas, over water, in five wide-mouthed 32-oz. bottles. Prepare a deflagrating-spoon by slipping the wire handle through a hole in the centre of a piece of sheet tin, larger than the mouths of the bottles (Fig. 27). Remove one of the bottles from the water by placing a glass plate over the mouth, holding it in position, and raising plate and bottle from the water. Place the bottle upright, and insert a glowing pine splinter. Into a second bottle lower a deflagrating-spoon on which is a small piece of ignited sulphur. Repeat with a piece of phosphorus the size of a pea. Test the water remaining in the jars with blue litmus paper. Repeat, using a small piece of carbon heated to redness.

Heat in a Bunsen flame a 4-inch piece of steel watch-spring, and straighten it. Slip one end of the spring through the sheet-tin cover used with the deflagrating-spoon, and fasten with a cork. Wrap a bit of cotton round the other end, and dip it into a little molten sulphur. Remove a jar of oxygen from the water, light the sulphur, and when it is burning plunge the watch-spring into the oxygen.

What is the difference between the above combustions and combustions in the air? Why?

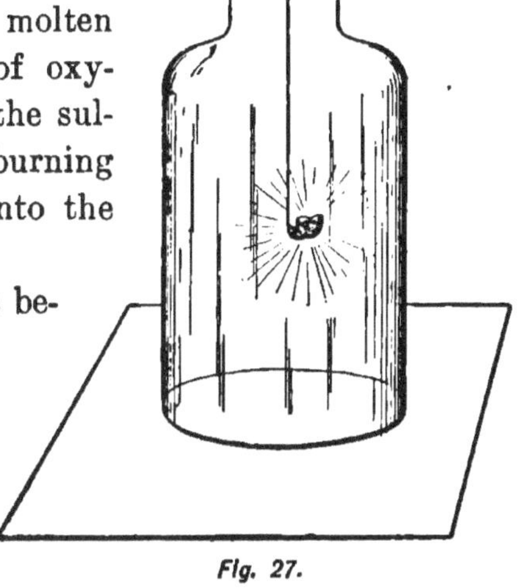

Fig. 27.

41. Reversing the Phenomena of Combustion. Burning of Oxygen in Hydrogen.

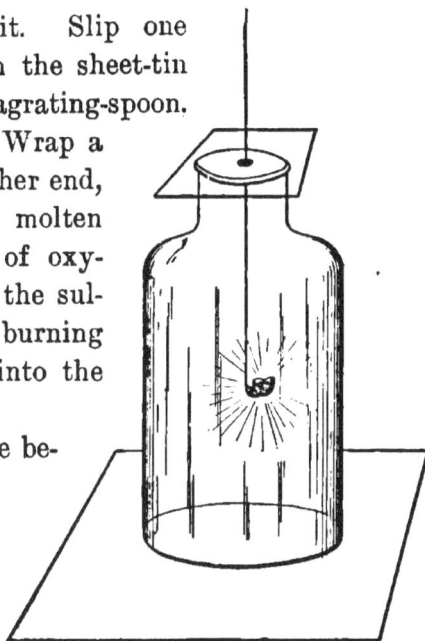

The apparatus necessary to demonstrate the combustion of oxygen in hydrogen is shown by Fig. 28. *C* is the neck of an ordinary retort. Into the narrow end of this is fitted a single-bored rubber stopper, which connects at *A* with a generator furnishing a brisk current of dry hydrogen, generated as in Experiment 9*a*. Allow the current of hydrogen to pass for some time, until all air is expelled; then light the hydrogen at the wide end of *C*. Have ready a burner, *B*, with a platinum tip (described in Experiment 12), through which a slow stream of oxygen is passing from the gasometer. The oxygen is dried by means of a small tube, filled

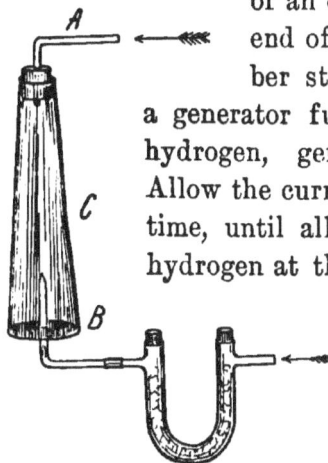

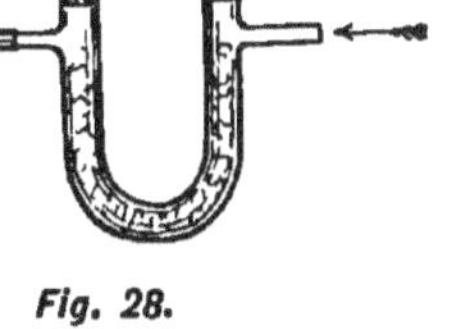

Fig. 28.

with anhydrous calcium chloride. (Wash-bottles cannot be used, because the bubbling of the gas through the liquid will cause the flame to flicker.) Thrust the burner, through which oxygen is passing, up into the tube *C*. As it passes through the burning hydrogen at the mouth, the oxygen will be ignited, and will continue to burn in *C*. What is the cause of the flame?

42. Heating of Iron Filings and Sulphur. Incandescence during Union.

Thoroughly mix in a mortar 7 grams of fine iron filings and 4 grams of flowers of sulphur. Place the mixture in a large test-tube, and hold it in the flame so that it is strongly heated at one point. When the mass begins to glow, remove the tube from the flame. What resemblance is there between this change and those encountered in ordinary combustion? What is produced?

43. *a*. Combustion of Phosphorus in Chlorine. (Experiment under Hood.)

Place a piece of phosphorus as large as a pea in a deflagrating-spoon, fitted with a piece of sheet tin as described in Experiment 40. Ignite the phosphorus, and lower it into a jar of dry chlorine. Compare this combustion with that of phosphorus in oxygen. What is the product in each case? Is there an evolution of light and heat?

***b*. Heat changes during the neutralization of acids by bases.**

Repeat Experiment 4*c*, using in the beaker 10 c.c. of sodium hydroxide solution, containing 5 grams of sodium hydroxide to 10 c.c. of water. Gradually add hydrochloric acid (one part concentrated hydrochloric acid to one part of water), stirring with the thermometer. Note change in temperature. Repeat, adding sulphuric acid (one part sulphuric acid to five parts of water).

44. Burning of Phosphorus and of a Candle in a Closed Air Space, and testing the Air which remains. Fig. 29.

a. Prepare a float made of a flat cork, on which is fastened a porcelain crucible cover. Place this cork in a pneumatic trough, with a piece of phosphorus the size of a bean on the cover, and light the phosphorus with a hot wire. Invert a bell jar of three litres capacity over the float, and slightly raise the stopper at the top, so as to let out the air, which expands greatly owing to the heat given off by the burning phosphorus. If this precaution is omitted, the air will be forced out at the bottom of the jar in large bubbles, and the disturbance may tip over the phosphorus float. After the violent combustion is over, insert the stopper of the bell jar, and allow to cool. To test the remaining gas, add enough water to the pneumatic trough to make the level within and without the bell jar alike, and then introduce a lighted taper.

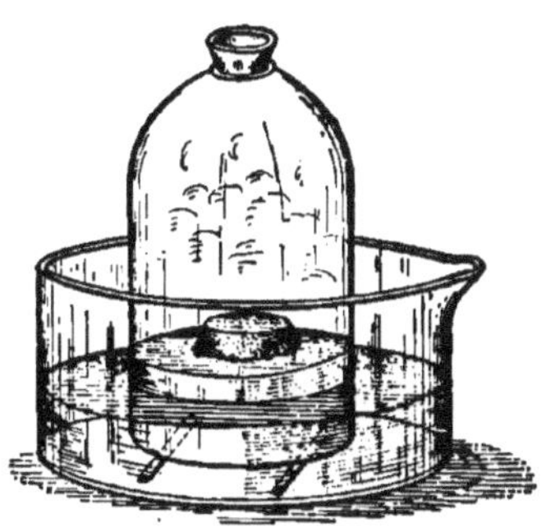

Fig. 29.

b. Repeat, attaching a piece of candle to a cork float, lighting the candle, and bringing it under the bell jar. When the light is extinguished, test the gas as before. Compare this with the action of air on copper. What is the distinction between combustion in air and in oxygen? Does the candle use up all of the oxygen before it is extinguished? What becomes of the phosphorus? Test the water in the apparatus with blue litmus.

45. Determination of the Volumetric Composition of the Atmosphere. Fig. 30.

Take a long glass tube, closed at one end, and divide it into five equal parts by means of rubber rings. Invert this over a long cylinder filled with water, so that the level without and within is at the first ring, and then clamp the tube in place. Fix a piece of phosphorus the size of a bean on a long copper wire, bend the wire as shown in the cut, thrust the phosphorus up into the tube, and set the apparatus aside for two days. Then sink the tube so that the level of the water without and within is the same. What is the ratio between the volume of gas remaining and the volume of gas originally present? By noting the height of the barometer before and after the experiment, and then applying the necessary corrections, quite accurate results can be obtained, if a carefully graduated tube is substituted for the crudely divided one indicated.

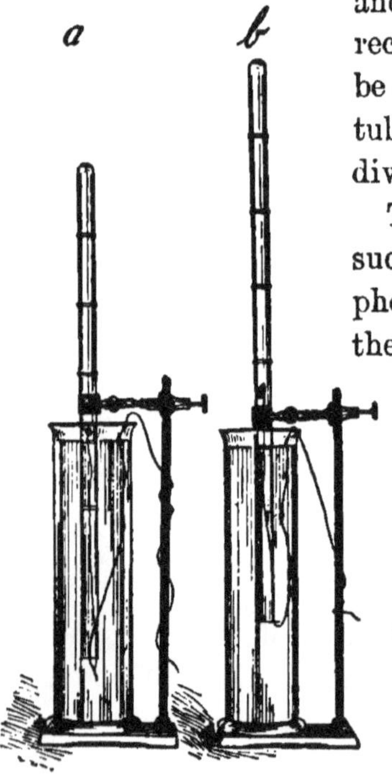

Fig. 30.

The figures (No. 30) show a tube such as is described. In one (*b*) the phosphorus has just been thrust into the enclosed air space; in the other (*a*) the same tube has been allowed to stand two days, and then has been adjusted so that the level without and within is the same.

46. Accurate Determination of the Volumetric Composition of the Atmosphere by Means of the Eudiometer Tube. Teacher's Experiment.

In order to measure accurately the relative amounts of oxygen and nitrogen in the atmosphere, the eudiometer is employed (Fig. 8, Experiment 10). The instrument should have a capacity of 100 c.c., should be partially filled with mercury, and inverted over the mercury trough so that about 25 c.c. of air will remain enclosed. About 14 c.c. of dry hydrogen are now run in, by slanting the tube and placing under its mouth the delivery tube of a hydrogen apparatus which is generating pure, dry hydrogen. Take all the precautions mentioned in Experiment 10, and ignite the mixture of gases with an electric spark. Be sure to read accurately the volume of air and the volume of hydrogen before the explosion, and also to measure the height of the column of mercury at each step, as indicated in Experiment 10. After the explosion, allow it to stand for ten minutes, and then read the volume of remaining gas, and reduce to standard conditions exactly as was done before. Note the diminution in volume. The hydrogen will have united with the oxygen to form water, therefore one-third of the diminution in volume must have been due to oxygen. Why? What is the volumetric composition of the atmosphere as given by the above experiment?

47. Determination of the Dew Point.

Suspend a thermometer over a beaker of water so that the bulb is immersed in the liquid. Note the temperature. Gradually add small pieces of ice to the water, and stir the liquid until a point is reached when moisture begins to collect on the outside of the beaker. Note the temperature.

48. Proof that Carbon Dioxide occurs in the Atmosphere.

a. Make a solution of calcium hydroxide (lime-water) as in Experiment 25*b*. Place this in a beaker, and expose it to the air for some time. What is observed? Explain.

b. Arrange an apparatus as in Experiment 39, Fig. 26, substituting an empty bottle in place of the tube containing copper shavings. Fit this bottle with a double-bored rubber stopper, so arranged that air can be sucked into this empty bottle through two wash-bottles containing sodium hydroxide. When all connections are made, open the siphon, and draw air slowly through the apparatus.

Remove the empty bottle, pour quickly into it some clear lime-water, and tightly stopper. Does the lime-water change? What alteration was produced in the air by sucking it through the caustic soda solution? After the lime-water has stood for some time in the empty bottle, open the bottle, introduce into it a small piece of glowing charcoal, and stopper again. Now observe the lime-water.* Take a little clear lime-water in a test-tube, and blow a current of air through it by means of a glass tube. What change do you observe?

49. Preparation of Ammonia Gas from a Solution of Ammonia in Water. Fig. 31. (Experiment under Hood.)

In a generating-flask, *a*, of 500 c.c. capacity place 200 c.c. of strong ammonia water (*aqua ammonia fortior*). Connect the delivery tube with a so-called drying-tower, *b*, which is a cylinder with a tubulated opening below, and with a single-bored stopper and delivery tube above. The drying-tower contains pieces of quicklime. When the apparatus is ready, heat the gas gently, and collect several cylinders of it by upward displacement of the air, *c*. Finally, collect the remaining gas in a beaker of water. Cover the jars with glass plates as fast as they are filled.

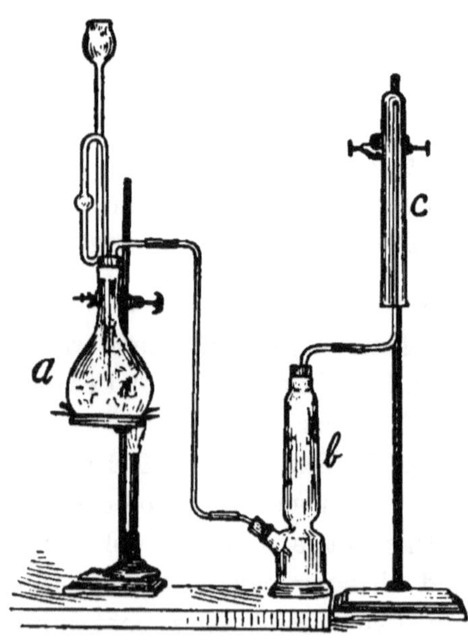

Fig. 31.

* Charcoal, in burning, produces carbon dioxide.

50. Experiments with Ammonia.

a. Place one jar, mouth downward, in a vessel of water. Is ammonia very soluble in water? Introduce a lighted splinter into the second jar. Test the odor of ammonia as you did that of chlorine. Test the solution of ammonia in water by means of a piece of red litmus paper. Afterwards expose this litmus paper to the air for a short time.

b. Bring together, mouth to mouth, a jar of ammonia and an equal-sized jar of oxygen. Tightly hold the two in position, and invert them several times. Finally apply a lighted splinter to the mouth of each jar.

51. Decomposition of Ammonia by Metals. Fig. 32.

Arrange an apparatus for the preparation of dry ammonia, as in Experiment 49. Connect the delivery-tube with a tube of infusible glass, as in Experiment 39. Fill the bulb of this tube with shavings of magnesium; connect the farther end with a safety-bottle, and from the latter run a delivery-tube to a vessel of water. Invert a test-tube full of water in the vessel, and have it ready for collecting the gas evolved. When all is ready, heat the ammonia water in the generating-flask very gently, so as to secure a

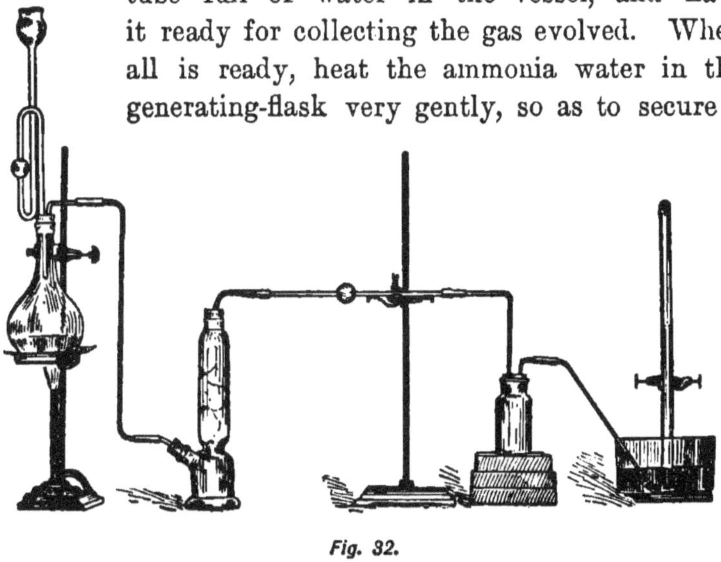

Fig. 32.

very slow evolution of gas. When all air has been expelled from the apparatus, heat the bulb containing magnesium to a low red heat, and collect the evolved gas in the inverted test-tube. Remove the test-tube, closing its mouth with the thumb, and instantly apply a lighted taper. What change in the appearance of the magnesium? Expel all ammonia from the bulb-tube by means of a current of air, and add water to the contents of the bulb. What odor do you note?

52. Decomposition of Ammonia by Chlorine.

Prepare a saturated solution of common salt, invert over it a 32-oz. bottle filled with the same liquid, and then replace this liquid by chlorine. Next, cover the chlorine bottle with a glass plate, and transfer it, mouth downward, to a vessel containing ammonia water. Does the color of the chlorine disappear? Does the gas diminish in volume? When no further change takes place, again cover the bottle with a glass plate, transfer, mouth downward, to another vessel containing dilute sulphuric acid (1 part sulphuric acid to 20 of water), and allow it to stand a few minutes. Then cover again, invert the bottle, and introduce a burning pine splinter. Has the odor of the chlorine disappeared? What is the gas remaining?

Fig. 33.

53. Determination of the Volumes of Hydrogen and Nitrogen in a Known Volume of Ammonia Gas. Fig. 33.

Take a tube one-half metre in length and 15 cm. in diameter, closed at one end. Divide this tube into three equal parts by two rubber rings. Fill it with a saturated solution of common salt, and invert it over a vessel containing the same liquid. Now expel the salt solution by means of chlorine. When the tube is entirely filled with chlorine, close the mouth with the thumb, and transfer, mouth downward, to a vessel containing a solution of ammonia in water. When action has ceased transfer again, keeping the mouth tightly closed with the thumb, to a cylinder one-half metre in depth, which is filled with a dilute solution of sulphuric acid (1 part of sulphuric acid to 20 of water). Adjust the tube so that the level of the liquid without and within is the same. At what point does it now stand? Test the gas which remains as in Experiment 52.

The tube was divided into three equal parts, and filled with chlorine. When the chlorine acted on the ammonia dissolved in water, it decomposed an equivalent quantity of ammonia, and set free a corresponding amount of nitrogen. This nitrogen occupies one-third the volume of the chlorine used; but the chlorine has united with the hydrogen which was originally combined with this one-third volume of nitrogen. As chlorine unites with hydrogen, volume for volume, to form hydrogen chloride (see Experiment 20), it follows that the three volumes of chlorine, indicated by the three divisions of the tube, must have united with three volumes of hydrogen; hence, 1 volume of nitrogen was united with 3 volumes of hydrogen in ammonia. The hydrochloric acid which was formed was removed from the tube by the ammonia water, and any ammonia which may have entered the tube has been removed by the dilute sulphuric acid, so that finally nothing but pure nitrogen is left.

54. Decomposition of Ammonia by the Electric Spark. Teacher's Experiment.

Fill a eudiometer tube with clean, dry mercury, and invert it in a mercury trough. Introduce 15 c.c. of dry ammonia gas prepared as in Experiment 49. Note accurately:—

1. Volume of gas.
2. Height of barometer.

3. Temperature.
4. Height of column of mercury in tube.

Connect the two platinum wires with an induction coil, and allow electric sparks to pass through the ammonia until no further increase in volume is observed; then disconnect the coil, allow to stand for five minutes, and note:—

1. Volume of gas.
2. Height of barometer.
3. Temperature.
4. Height of column of mercury in tube.

Reduce the gas volume to standard conditions of temperature and pressure. What is the relation between the first volume and the second? Combining what you learned in Experiments 53 and 54, what volumes of nitrogen and hydrogen unite to form ammonia? What volume of ammonia is produced from these volumes of hydrogen and nitrogen? Compare the results with the ones obtained in determining the volumetric composition of water and of hydrogen chloride. In the formation of hydrogen chloride, 1 volume of hydrogen + 1 volume of chlorine = 2 volumes of hydrogen chloride. In the formation of water, 2 volumes of hydrogen + 1 volume of oxygen = 2 volumes of water vapor. What is the case with ammonia, as shown by Experiment 54?

55. Action of Ammonia on Hydrogen Chloride.

A cheap apparatus, which will demonstrate that equal volumes of hydrogen chloride and ammonia unite to produce a solid, is described in the text, page 98. This experiment can, however, be performed more accurately and neatly by using the apparatus described in Experiment 20, Figs. 16 and 17.*

* The glass stopcock must have a very large bore, otherwise the ammonium chloride will fill it, and prevent communication between the tubes.

Manipulation exactly as in the demonstration that hydrogen chloride is formed of equal volumes of hydrogen and chlorine. Through the three-way stopcock (Fig. 17) pass a current of dry ammonia (Experiment 49, Fig. 31), having turned the stopcock so that the gas passes in through one of the tubes, and out at the narrow tip. Allow the current to pass until you are sure that all the air is expelled (five to ten minutes). Now close this arm with the stopcock, and seal the tip in the flame of the Bunsen burner. Next, fill the second arm in the same way with dry hydrogen chloride (Experiment 16, Fig. 13), and, after sealing the tip, turn the stopcock so that the two tubes communicate. Allow the apparatus to stand for twenty-four hours, and then, after scratching one of the tips with a file, dip it under dry mercury, and break the tip.

56. Preparation of Ammonium Chloride. The Decomposition of Ammonium Chloride by Slaked Lime.

a. In a beaker place 25 c.c. of ammonia solution. Take out a drop, and place it on a strip of red litmus paper. Allow the litmus paper to stand in the air for some minutes, and then examine. To the solution of ammonia add, gradually, a solution of hydrochloric acid until neutral toward litmus. Now evaporate the solution to dryness in a porcelain dish on a water-bath. What is the appearance of the remainder? Heat a little on a piece of platinum foil in the Bunsen burner. What is the effect?

b. In a porcelain evaporating-dish place about 10 grams of quicklime, and gradually, a few drops at a time, add water. Keep on adding until the lime is of the consistency of very thick paste. Now arrange an apparatus as shown in Fig. 31, Experiment 49; place the slaked lime in the generating-flask, and add to it 10 grams of your ammonium chloride, which you have powdered in a mortar.

Connect all parts of the apparatus, and heat the generating-flask very gently. Collect the gas which passes off by displacement of air, as was done in Experiment 50. What gas has been collected? Identify it by its action on a piece of moist red litmus paper, and by its solubility in water. Over the mouth of one of the tubes of gas hold a glass rod, on the tip of which is a drop of a concentrated solution of hydrochloric acid in water. What is the effect?

57. Formation of the Ammonium Salts of Various Acids. The Decomposition of these Salts by Bases.

a. In a beaker place 5 c.c. of a diluted solution of nitric acid (1 : 5), and then add ammonia solution until neutral toward litmus. What is the change in temperature? Evaporate to dryness on a water-bath. Heat a little of the salt thus produced on a piece of platinum foil in the Bunsen burner. To some of the salt in a test-tube add a concentrated solution of potassium hydroxide (one part caustic potash to one of water). Odor of gas passing off? In the mouth of this test-tube suspend a moist strip of red litmus paper, so that it does not touch at the sides. Over the mouth of this test-tube hold a glass rod, on the tip of which is a drop of a concentrated solution of hydrochloric acid.

b. Repeat, using diluted sulphuric acid (1 : 5) instead of nitric acid.

c. Repeat *a* and *b*, using sodium hydroxide instead of potassium hydroxide.

58. The Determination of the Volumes of Hydrogen produced by the dissolving of Known Weights of Metals in Acids. (Repeat Each of the Following Experiments until the Two Results in Each Case exactly agree.)

a. Arrange an apparatus as shown in Fig. 34. It consists of a glass beaker, *a*, into which is placed a small funnel, *c*, the stem of which is cut off. Accurately weigh off about .1 gram of pure zinc, place it under the inverted funnel, and nearly fill the beaker with distilled water. Now fill a gas-measuring tube, *b* (which has a capacity of 200 c.c., and which is graduated in $\frac{1}{5}$ c.c.), with diluted hydrochloric acid, invert the tube in the beaker, and fix it in position, so that its mouth is over the opening of the funnel. Owing to its greater specific gravity, the hydrochloric acid will descend into the water, and will come in contact with the zinc. All of the hydrogen which passes off must necessarily collect in the gas-measuring tube. When all of the zinc has disappeared, tightly close the tube with the thumb, and transfer it to a deep cylinder of pure water; clamp so that the level without and within is the same. Note temperature and barometer, and recalculate the gas-volume to standard conditions. What weight of hydrogen, in grams, does this volume represent? From the data at hand ascertain the quantity of zinc which would be necessary to liberate one gram of hydrogen.

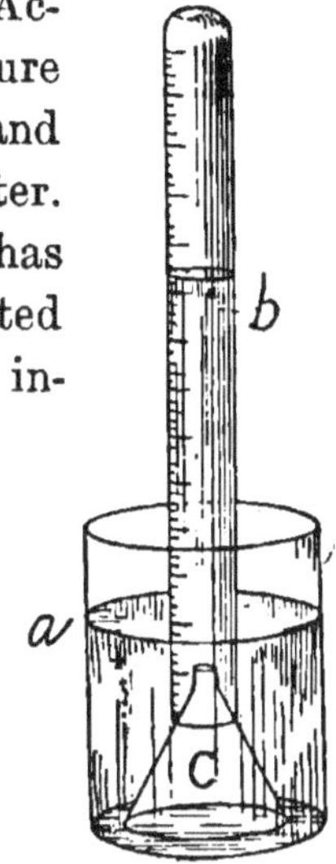

Fig. 34.

b. Repeat *a*, using dilute sulphuric acid (1 : 10) instead of hydrochloric acid.

c. Repeat *a* and *b*, using about .1 gram of pure iron (piano wire).

d. If desired, *a* and *b* can be repeated, using a piece of magnesium ribbon which has been carefully cleaned. In this case a gas-measuring vessel of at least 300 cubic centimetres capacity must be selected. What are the equivalent weights of iron and zinc as shown by the experiments?

59. Alteration in Wood by Heating.

In a test-tube of so-called infusible glass, place a piece of fresh pine wood, and bring it into the flame of the Bunsen burner. Gradually increase the heat. Alteration in the appearance of the wood? Odor? Apply a lighted taper to the mouth of the tube.

60. *a.* The Absorption of Coloring Matter by Charcoal.

1. In a test-tube place a little indigo, add to this a few drops of concentrated sulphuric acid, warm slightly, and then dilute with water, and filter. Pour the blue solution into a 300 c.c. flask in which you have placed 10 grams of fresh animal charcoal.* Warm the solution, and shake vigorously for a short time; then filter off the charcoal. What is the color of the filtrate?
2. Repeat, using a solution of blue litmus.
3. Repeat, using a solution of logwood.

b. Absorption of gases by charcoal.

Arrange an apparatus for the preparation of ammonia as in Experiment 49, Fig. 31. Collect a test-tube full of the gas over dry mercury, and leave the test-tube clamped in position, so that its mouth dips under the mercury. Now heat a small piece of wood-charcoal for an instant in the flame of the Bunsen burner.† After allowing any portion which may be glowing to become extinguished, slip it quickly under the mercury, so that it will rise in the test-tube. When no further absorption takes place, prepare a second. piece of charcoal as above, and bring it into the tube. What is true of ammonia gas is true of a large number of other gases which, like ammonia, are easily condensed to liquids (sulphur dioxide, hydrogen sulphide, etc.).

* Animal charcoal is to be obtained from any dealer in chemicals.
† Hold the charcoal by a pair of pincers or a pair of tongs.

61. Burning of Charcoal in a Closed Space of Oxygen.

Prepare a glass tube 30 cm. in length and 10 cm. internal diameter, sealed at one end, and bent as shown in Fig. 35. Slip into the tube a piece of wood-charcoal the size of a pea, fill the tube with mercury, and invert it in a small vessel, *b*, of the same liquid, clamping it in position as shown in the cut. Now admit dry oxygen into the tube until the mercury is expelled as far as the bend, and mark the position of the liquid by means of a strip of paper. Now gently heat the charcoal with the Bunsen burner until combustion begins, then remove the flame, and allow the charcoal to stand until combustion is complete, and the apparatus has cooled to its former temperature. Is there any change in the volume of the gas?

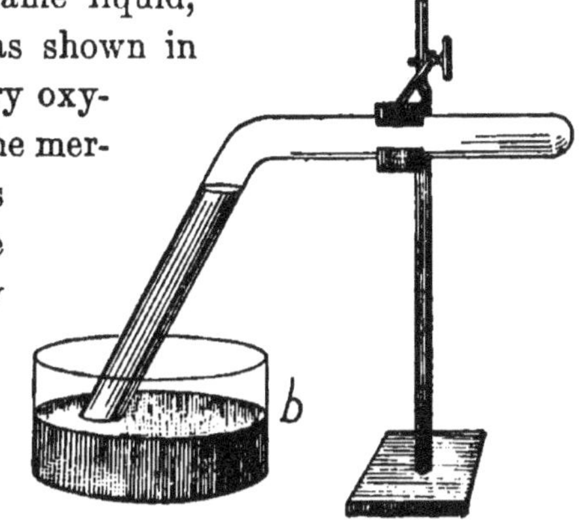

Fig. 35.

62. Formation of Carbon Dioxide by passing Oxygen over Hot Charcoal.

Arrange the apparatus used in Experiment 21, Fig. 18, with the difference that in place of the hydrochloric acid generator (*A*), you have an apparatus delivering dry oxygen. Fill the iron tube with about 15 grams of charcoal in pieces the size of a pea, connect all parts of the apparatus, and collect the gas which passes off by displacement of air in several cylinders placed mouth upward. Cover each cylinder, after the air is expelled, with a glass plate. Into one of the cylinders introduce a burning taper. Into another pour a few drops of a filtered solution of slaked lime

[Experiment 25*b*] (formation of calcium carbonate). Invert a cylinder in a vessel containing a solution of caustic potash. In one of caustic soda.

63. Formation of the Primary and Secondary Carbonates of Potassium and Sodium.

For these experiments, arrange an apparatus for generating carbon dioxide as described in Experiment 65.

a. Prepare a solution of 10 grams of potassium hydroxide in 50 grams of water, place this solution in a small beaker, attach a funnel small enough to pass inside this beaker to the exit tube of an apparatus generating carbon dioxide, and then dip the funnel into the beaker so that its edge is just below the surface of the liquid (Fig. 36). Now pass a brisk current of carbon dioxide into the solution of caustic potash until a drop, taken out and placed on a piece of red litmus paper, is neutral. Now evaporate gently to dryness in a porcelain dish on a water-bath. What is the appearance of the remainder? Is it like caustic potash? Place a little of the remainder in a test-tube, and add to it a few drops of diluted hydrochloric acid, and then in the mouth of the tube hold a glass rod, on the tip of which you have a drop of clear solution of slaked lime. Compare this behavior with that of caustic potash under the same conditions. Heat some of the remainder in a test-tube, and hold over the mouth a glass rod, on the tip of which is a drop of a clear solution of slaked lime.

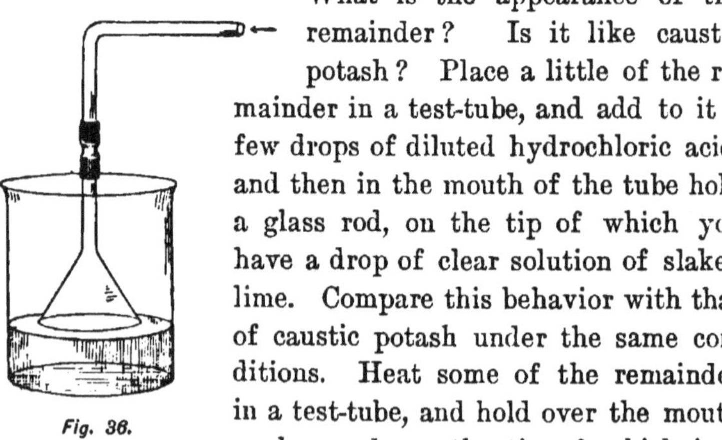

Fig. 36.

b. Purify 5 grams of the primary carbonate of potassium (prepared by the above experiment), by dissolving it in cold distilled water, evaporating the solution at a gentle heat to crystallization, and then drying the crystals between pieces of filter paper. Finally, powder the crystals, and place them in a desiccator over concentrated sulphuric acid for twenty-four hours. Take a gas-measuring tube (of 100 c.c. capacity, and graduated to .1 c.c.), fill it with dry mercury, and invert it over a trough of the same liquid. Now carefully weigh off .2 grams of your pure primary carbonate of potassium, wrap it in a piece of filter paper so as to form a package small enough to slip up into the mercury tube, and then bring it under the mouth so that it will rise to the top. From a curved pipette (Fig. 24) introduce a few drops of diluted hydrochloric acid into the tube. What is the effect? When the gas no longer increases in volume, introduce another drop of the acid, and see if it has any further effect. Carefully continue this operation until no further evolution of gas and increase in volume are observed. Now allow the tube to stand for some minutes, and then measure the gas volume, taking account of the temperature, barometer, and height of the column of mercury in the gas-measuring tube. Reduce the gas volume to standard conditions of temperature and pressure. What weight of carbon dioxide is liberated by one gram of potassium primary carbonate? *

c. Carefully weigh off 2 grams of pure, dry primary carbonate of potassium, and place it in a deep porcelain crucible or test-tube of infusible glass. Weigh the whole, and then heat, gently at first, and finally to a red heat for fifteen minutes. Allow it to cool, and again weigh. What is the loss in weight? What is the weight of the remainder? Continue the above operation until no further change in weight takes place. What volume of carbon dioxide would this loss in weight represent? Compare

*** One cubic centimetre of carbon dioxide weighs .001986 gram.**

this volume with that given off by 2 grams of primary potassium carbonate when treated with acids (*b*). Weigh carefully .138 gram of pure, dry secondary potassium carbonate, and with this repeat *b*. Compare the volume of carbon dioxide given off with that formed in *b*. From the data thus obtained, calculate the weight of primary potassium carbonate which would contain one molecular weight of carbon dioxide. Calculate the amount of potassium hydroxide combined with one molecular weight of carbon dioxide.

64. Prepare a solution of potassium hydroxide as indicated in Experiment 27*a*. Now weigh off 1 gram of primary potassium carbonate, dissolve in 25 c.c. of water, place in a beaker, and carefully add to it from the burette an amount of potassium hydroxide solution representing .56 gram of solid potassium hydroxide. Evaporate the solution so formed to dryness, and heat a little of the remainder in a test-tube. Does it give off carbon dioxide? To a little in a test-tube add a few drops of a dilute solution of hydrochloric acid. Hold a glass rod, on the tip of which is a drop of a clear solution of slaked lime, over the mouth of the tube, after adding acid. Is the product of the action of potassium hydroxide on primary potassium carbonate primary or secondary potassium carbonate?

65. Preparation of Carbon Dioxide for Laboratory Use.

The apparatus is the same as that represented by Fig. 13.

In the generating-flask place about 25 grams of clean marble * in pieces the size of a hickory-nut. Connect all parts of the apparatus (have some pure water in the washing-bottle), and then pour diluted hydrochloric acid (one

* Completely fill the flask with water before dropping in the pieces of marble, and after all is added pour off the water. If this precaution is not taken, the pieces of marble may crack the flask as they are dropped.

part concentrated acid to five of water) in through the safety tube. Collect the gas which is evolved in a cylinder, which is placed mouth upward. Place a burning candle in a cylinder of air, and then pour into it carbon dioxide from a second cylinder, just as one would pour water. Into a cylinder of carbon dioxide place a little blue litmus solution, cover the cylinder with a glass plate, and shake. Invert a test-tube full of carbon dioxide over a solution of potassium hydroxide.

66. Formation of the Secondary and Primary Carbonates of Calcium.

In a beaker place 15 c.c. of a clear solution of slaked lime. Into this solution pass carbon dioxide from the generator. Continue to pass in carbon dioxide until the white precipitate, which is at first produced, disappears. When the solution is clear, discontinue the carbon dioxide, and then warm the water in the beaker to boiling.

67. Preparation of Carbon Monoxide.

Arrange a carbon dioxide apparatus as in Experiment 66. Connect the delivery tube with an iron tube which you can heat in a furnace, as in Experiment 21, Fig. 18. Introduce a safety-bottle, *c*, after the iron tube; put in the tube 15 grams of charcoal in pieces the size of a pea, connect all parts of the apparatus, heat the iron tube to redness, and then pass a slow current of carbon dioxide over the charcoal. Collect the evolved gas in cylinder over a dilute solution of potassium hydroxide (1 : 20). When the quantity is sufficient, remove one of the cylinders, place it mouth upward, and introduce a lighted taper. After combustion is over, pour a little of a clear solution of slaked lime into the cylinder, cover with a glass plate, and shake. Is the gas which you have obtained by the action of carbon dioxide on charcoal soluble in potassium hydroxide? How does it differ from carbon dioxide?

68. Preparation of Methane. (Hydrogen Carbide.)

The apparatus is shown by Fig. 37. The tube *a* is of so-called infusible glass, sealed at one end, and stoppered with a single-bored rubber stopper, which is connected with a safety bottle and delivery tube. Prepare an intimate mixture in a mortar, of 10 grams of sodium acetate (which has previously been dried by heating to fusion in an open dish), and 10 grams of a mixture of sodium hydrox-

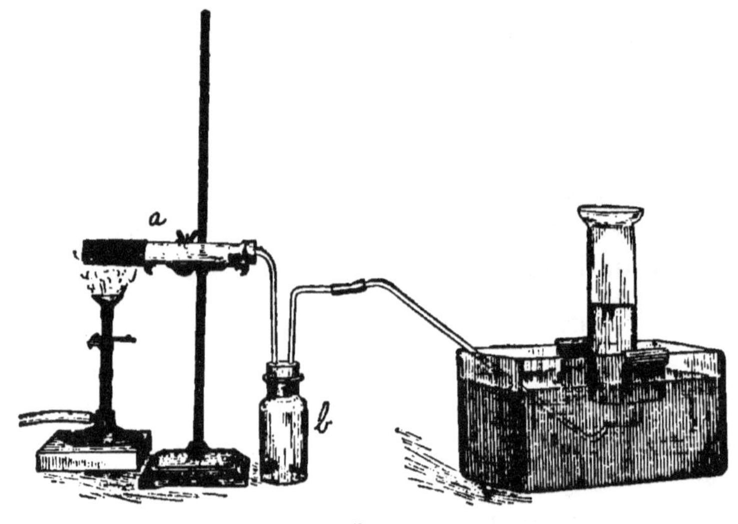

Fig. 37.

ide and calcium oxide (so-called soda-lime),* and then half fill your tube with this mixture. Place the tube on its side, and pound it lightly on the table until you have formed a canal for the free passage of the gas.† Now connect all parts of the apparatus, and wrap a piece of wire gauze around the tube, to prevent its cracking when

* Soda-lime can be purchased of dealers in chemicals.

† One must be absolutely certain that there is no packing of the mixture in the tube. Any portion of the tube which may be closed gas-tight will cause a dangerous explosion, by preventing the escape of gas.

heated. Then heat gently with a Bunsen burner, beginning at the closed portion of the tube, until there is a regular evolution of gas. Wait until all air is expelled from the apparatus, and then collect in cylinders or test-tubes, over water.

69. Properties of Methane

Remove one of the cylinders from the water, invert it, and instantly apply a lighted taper to the mouth. Remove a second cylinder, place it mouth upward, allow it to stand for a few seconds, and then apply a lighted taper. What is the effect? Is methane specifically heavier or lighter than air? Remove a third cylinder, light the gas while it is held mouth downward; now invert the cylinder. Into a cylinder in which you have burned methane pour a few drops of a clear solution of slaked lime, cover the cylinder with a glass plate, and shake.

70. The Volumetric Composition of Methane. Teacher's Experiment.

The apparatus is given by Fig. 9, Experiment 11. Fill the graduated eudiometer tube with clean, dry mercury, and invert it in the deep cylinder as described in Experiment 11. Now introduce about 10 to 15 c.c. of pure, dry methane (prepared as in Experiment 68) into the tube, pass steam into the jacket until the temperature is constant, and then carefully measure the gas-volume after taking the usual observations. (What are they?) Now pass in approximately 25 c.c. of dry oxygen, allow to remain for a minute, and again measure. Then lower the tube into the mercury cylinder as far as it will go, wrap a towel around the mouth of the cylinder, so as to prevent spattering, and explode with a spark from an induction coil (Experiment 10).* Bring the eudiometer tube back

* Be sure that the steam-jacket is thoroughly heated during the entire experiment.

into position, so that the height of the column of mercury is the same as it was before the explosion, and note if there was any change in the volume of gas. Now allow to cool until the entire apparatus is at the temperature of the surrounding air; then introduce a drop of water into the eudiometer,* and measure the volume as before, this time taking the tension of water vapor into account during your calculations. Next, cut a *small* piece of potassium hydroxide, take it by a pair of pincers, and slip it under the eudiometer tube, so that it will rise to the upper level of the mercury. Allow to stand until no further contraction in volume takes place, and then again measure the gas-volume. (This time neglect the pressure of water vapor.) The volume of contraction on cooling corresponds to the volume of what product of the explosion? The volume of contraction after adding potassium hydroxide? From the data you have obtained, calculate the composition of methane by volume.

71. The Action of Chlorine on Methane.

Take two jars of the same size, with ground tops; fill one with chlorine (Experiment 21*b*), and the other with methane over water (Experiment 68). Cover each jar with a glass plate, bring them together, mouth to mouth, then remove the plates which separate them, hold tightly in position, and invert several times. When the gases have completely mixed, apply a lighted taper to the mouth of each. Result?

72. The Gradual Action of Chlorine on Methane.

a. The apparatus is that described in Experiment 20, Figs. 16 and 17, and the manipulation is the same, except-

* This drop of water is introduced so as to be certain that the gases to be measured are *completely* saturated with moisture.

ing that one of the arms of the double tube is filled with methane instead of hydrogen. The apparatus must stand for at least twenty-four hours in daylight or sunlight before the reaction is completed and the color of chlorine has entirely disappeared. After this time bring one of the arms of the tube into a solution of blue litmus, and, after scratching with a file, break the tip. Change in the color of litmus? What portion of the apparatus becomes filled with the litmus solution? After allowing to stand for a few minutes, remove the tube from the liquid, turn the stopcock so that this fluid can drain off, then bring the opened end of the tube into a deep cylinder of mercury (described in Experiment 11, Fig. 9), and press down until the liquid rises to the stopcock. Turn the stopcock as it was before, and then break off the second tube which extends upward, quickly apply a lighted taper to its mouth, and force out the remaining gas by pressing the apparatus down into the mercury. Color of flame? To the solution of blue litmus add a few drops of a solution of silver nitrate. Result? What insoluble substance separates?

73. The Action of Chlorine on Methyl Chloride.

Repeat Experiment 72, with the difference that after you have opened the tube under litmus solution you allow the liquid to drain off, and then, after drying, once more fill it with chlorine, and seal the tip, while the branch containing methyl chloride is carefully kept closed. After establishing communication between the two arms of the tube, allow to stand in the daylight, and then follow directions as in Experiment 72. Compare the two results.

74. The Action of Chlorine on Methylene Chloride.

Repeat Experiment 73, using the apparatus in which you have produced methylene chloride; i.e., which contains methane which has twice been acted on by chlorine. After the chlorine has acted on the methylene chloride for twenty-four hours, open the tube, as before, under a solution of litmus, and note the result. How does this differ from the two preceding cases? Odor of product which is formed?

75. The Formation of Salts by Double Decomposition.

a. To a solution of barium chloride in a test-tube add a few drops of a diluted solution of sulphuric acid. What is formed? Does the precipitate dissolve on boiling? Does it dissolve in an excess of sulphuric acid?

b. Repeat *a*, using a solution of potassium sulphate instead of one of sulphuric acid.

c. Repeat *b*, using a solution of magnesium sulphate.

76. Precipitation of Salts by Double Decomposition.

a. To a solution of calcium chloride add a few drops of a solution of potassium carbonate. What is formed? Does the precipitate dissolve on boiling? Filter the precipitate after boiling, and to the substance on the filter paper add a few drops of hydrochloric acid. Does the precipitate dissolve? See if the precipitate will dissolve in nitric acid. In sulphuric acid. What gas passes off in each case?

b. Repeat *a*, using a solution of barium chloride instead of one of calcium chloride.

c. To a solution of sodium chloride add a few drops of a solution of silver nitrate. What is formed? Does the precipitate dissolve on boiling? Does it dissolve in hydrochloric acid? In nitric acid? To the precipitate add a slight excess of a solution of ammonia in water. Does it dissolve?

d. Repeat *c*, using solution of hydrochloric acid instead of sodium chloride.

e. To a solution of magnesium sulphate add a few drops

of a solution of barium chloride, and repeat directions in Experiment 75*a*. Compare the result with that obtained in 75*a*.

77. The Action of Chlorine on Hydrobromic Acid, Hydriodic Acid, Sodium Bromide, and Sodium Iodide.

a. Obtain a solution of hydrobromic acid from the side-table, place a little in a test-tube, add to it three or four drops of carbon bisulphide, and then add a little of a solution of chlorine in water. Close the mouth of the tube with the thumb, and shake. Change of color in the carbon bisulphide? * What change takes place?

b. Repeat *a*, using a solution of hydriodic acid. (Take care not to add more than a little chlorine water, otherwise the color of the separated iodine may be destroyed.) Color of a solution of iodine in carbon bisulphide?

c. Repeat *b*, using a solution of bromine in water instead of one of cholorine.

d. Repeat *a*, *b*, and *c*, using solutions of potassium bromide and iodide instead of hydrobromic acid and hydroiodic acid.

78. The Preparation of Hydrogen Sulphide. (All Experiments with Hydrogen Sulphide must be performed under the Hood.)

The apparatus is the same as that used for the preparation of hydrogen chloride (Experiment 16, Fig. 13). Into the gas-generating flask place 10 grams of fused ferrous sulphide (precautions in this case as in Experiment 65) about the size of a bean, connect all parts of the apparatus, and add diluted sulphuric acid (1 : 5) through the safety tube. The wash-bottle should contain pure water. Collect

* Carbon bisulphide dissolves bromine and iodine so readily that it will extract them from their solutions in water. Their color is therefore rendered more apparent by being concentrated in the few drops of carbon bisulphide.

the gas in jars by displacement of air, taking care during the operation to partly cover each jar with a glass plate, so as to just leave room for the delivery-tube of the apparatus. Completely cover the jars with glass plates as fast as they are filled. Remove the cover of one of the jars, and apply a lighted taper. What separates on the sides of the jar? Why? Is the result different if you mix hydrogen sulphide with oxygen, and then ignite? (Perform this latter experiment in a test-tube, and wrap a towel around it before approaching the flame.) Odor of hydrogen sulphide? (Precautions as in testing the odor of chlorine.) Solubility of hydrogen sulphide in water?

79. The Action of Hydrogen Sulphide on the Salts of Metals. Formation of Sulphides which do not dissolve in Dilute Acids.*

a. In a test-tube prepare a solution of copper sulphate in water, add two or three drops of hydrochloric acid, and then pass hydrogen sulphide into the solution. Color of precipitate? Filter the precipitate, wash it with warm water, dry it on its filter, and then bring it into a second test-tube. Is it soluble in cold hydrochloric acid? In hot hydrochloric acid? In nitric acid? In a mixture of one part nitric acid to three parts hydrochloric acid?

b. Repeat *a*, substituting a solution of lead acetate for copper sulphate, and in this case acidify with two drops of nitric acid instead of hydrochloric acid.

c. Repeat *a*, using a solution of the oxide of arsenic in an excess of hydrochloric acid.

d. Repeat *a*, using a solution of silver nitrate, acidified with a few drops of nitric acid.

* In working with hydrogen sulphide, pass the gas into the solutions through a tube with a very small bore (a so-called capillary tube). By this means excessive waste of the gas by reason of a too rapid current is avoided, and the danger of poisonous effects is reduced to a minimum.

80. Formation of Sulphides which dissolve in Dilute Acids.

a. Repeat 79*a*, using a solution of zinc sulphate which you have *not* acidified. Filter whatever precipitate you may have, and test the clear filtrate with a strip of blue litmus paper. Is it acid? To the clear filtrate add a few drops of a solution of potassium sulphide.* Does zinc sulphide separate from this filtrate? To zinc sulphide add a few drops of hydrochloric acid. Odor of gas passing off on acidifying? Compare the result with that obtained by adding hydrochloric acid to calcium carbonate (Experiment 76).

To a solution of zinc sulphate which has been acidified with a few drops of sulphuric acid add hydrogen sulphide. Is there any precipitate? To this solution add potassium sulphide. Result?

b. Repeat *a*, using a solution of ferrous sulphate (green vitriol).

81. The Separation of Two Metals by using Hydrogen Sulphide.

In a test-tube place a little of a solution of zinc sulphate, and add to it the same quantity of a solution of copper sulphate. Acidify with three or four drops of dilute sulphuric acid, and then pass in hydrogen sulphide. Continue to pass in the gas until its odor remains in the solution after standing for some time. Now filter off the copper sulphide which is formed, and to the filtrate add a solution of potassium sulphide. Result?

* In order to prepare potassium sulphide, dissolve 5 grams of potassium hydroxide in 50 c.c. of water, take one-half of the solution, and pass in hydrogen sulphide until no more gas will dissolve. Now pour into this the other half of the solution.

The Principal Elements, with Their Atomic Weights.

Name of Element.	Chemical Symbol.	Usual Condition.	Atomic Weight.	Specific Gravity.
Aluminium	Al.	White, metallic solid	27.	2.56
Antimony	Sb.	Crystalline, metallic solid	120.	6.7
Arsenic	As.	Gray, crystalline solid	75.	5.7
Barium	Ba.	Reddish metal, attacked by water	137.5	3.7
Bismuth	Bi.	Reddish crystalline metal	208.9	9.7
Boron	B.	Grayish-black powder	11.	2.68
Bromine	Br.	Dark brown liquid	80.	3.2 *
Calcium	Ca.	Reddish metal. Attacks water	40.	1.57
Carbon	C.	Solid, appears in 3 forms: diamond, graphite, and charcoal	12.	2.2
Chlorine	Cl.	Greenish-yellow gas	35.5	2.45 †
Chromium	Cr.	Grayish metallic powder	52.1	6.8
Copper	Cu.	Red metal	63.6	8.8
Gold	Au.	Yellow metal	197.3	19.3
Hydrogen	H.	Colorless gas	1.	.069 †
Iodine	I.	Black, crystalline solid	127.	4.95
Iron	Fe.	Grayish metal	56.	7.8
Lead	Pb.	Gray, soft metal	207.	11.44
Magnesium	Mg.	White metal	24.3	1.74
Manganese	Mn.	Grayish-white metal	55.	8.
Mercury	Hg.	White, liquid metal	200.	13.6
Nitrogen	N.	Colorless gas	14.	.972 †
Oxygen	O.	Colorless gas	16.	1.105 †
Phosphorus	P.	In 2 forms: a yellow solid and a red powder	31.	1.83 to 2.1
Potassium	K.	Grayish metallic solid. Attacks water	39.	.865
Silicon	Si.	Black solid in small, lustrous crystals	28.4	2.5
Silver	Ag.	White, metallic solid	108.	10.5
Sodium	Na.	White, metallic solid. Attacks water	23.	.972
Sulphur	S.	Yellow solid	32.	2.045
Tin	Sn.	White metal	119.	7.3
Zinc	Zn.	Grayish-white metal	65.3	7.15

* Specific gravity as liquids.

† Specific gravity as gases; air = 1.

Descriptive Inorganic General Chemistry

A text-book for colleges, by Professor PAUL C. FREER, University of Michigan. Revised Edition. 8vo, cloth, 559 pages. Price, $3.00.

IT aims to give a systematic course of Chemistry by stating certain initial principles, and connecting logically all the resultant phenomena. In this way the science of Chemistry appears, not as a series of disconnected facts, but as a harmonious and consistent whole.

The relationship of members of the same family of elements is made conspicuous, and resemblances between the different families are pointed out. The connection between reactions is dwelt upon, and where possible they are referred to certain principles which result from the nature of the component elements.

The frequent use of tables and of comparative summaries lessens the work of memorizing, and affords facilities for rapid reference to the usual constants, such as specific gravity, melting and boiling points, etc. These tables clearly show the relationship between the various elements and compounds, as well as the data which are necessary to emphasize this relationship. They also exhibit the structural connection between existing compounds.

Some descriptive portions of the work, which especially refer to technical subjects, have been revised by men who are actively engaged in those branches. In the Laboratory Appendix will be found a list of experiments, with descriptive matter, which materially aid in the comprehension of the text.

Professor Walter S. Haines, *Rush Medical College, Chicago.:* The work is worthy of the highest praise. The typography is excellent, the arrangement of the subjects admirable, the explanations full and clear, and facts and theories are brought down to the latest date. All things considered, I regard it as the best work on inorganic chemistry for somewhat advanced general students of the science with which I am acquainted.

Professor J. H. Long, *Northwestern University, Evanston, Ill.:* I have looked it over very carefully, as at first sight I was much pleased with both style and arrangement. Subsequent examination confirms the first opinion that we have here an excellent and a very useful text-book. It is a book which students can read with profit, as it is clear, systematic, and modern.

The Elements of Physics

By Professor HENRY S. CARHART, University of Michigan, and H. N. CHUTE, Ann Arbor High School. 12mo, cloth, 392 pages. Price, $1.20.

THIS is the freshest, clearest, and most practical manual on the subject. Facts have been presented before theories.

The experiments are simple, requiring inexpensive apparatus, and are such as will be easily understood and remembered.

Every experiment, definition, and statement is the result of practical experience in teaching classes of various grades.

The illustrations are numerous, and for the most part new, many having been photographed from the actual apparatus set up for the purpose.

Simple problems have been freely introduced, in the belief that in this way a pupil best grasps the application of a principle.

The basis of the whole book is the introductory statement that physics is the science of matter and energy, and that nothing can be learned of the physical world save by observation and experience, or by mathematical deductions from data so obtained. The authors do not believe that immature students can profitably be set to rediscover the laws of Nature at the beginning of their study of physics, but that they must first have a clearly defined idea of what they are doing, an outfit of principles and data to guide them, and a good degree of skill in conducting an investigation.

William H. Runyon, *Armour Institute, Chicago:* Carhart and Chute's text-book in Physics has been used in the Scientific Academy of Armour Institute during the past year, and will be retained next year. It has been found concise and scientific. We believe it to be the best book on the market for elementary work in the class-room.

Professor M. A. Brannon, *University of North Dakota, Grand Forks:* I am glad to express the opinion, based on the use of this work in Elementary Physics last year, at Fort Wayne, Ind., that it is the most logical and clear presentation of the subject with which I am acquainted. The problems associated with the discussion of Physical phenomena, laws, and experiments serve the dual purpose of leading the scholar to reason, and put into practice the previous clearly and concisely stated principles of Elementary Physics. It is a book that will greatly elevate the standard of scholarship wherever used.

www.ingramcontent.com/pod-product-compliance
Lightning Source LLC
Chambersburg PA
CBHW021510210326
41599CB00012B/1202